Estimation of Annual Agricultural Pesticide Use for Counties of the Conterminous United States, 1992–2009

By Gail P. Thelin and Wesley W. Stone

National Water-Quality Assessment Program

Scientific Investigations Report 2013–5009

U.S. Department of the Interior
U.S. Geological Survey

U.S. Department of the Interior
KEN SALAZAR, Secretary

U.S. Geological Survey
Suzette M. Kimball, Acting Director

U.S. Geological Survey, Reston, Virginia: 2013

For more information on the USGS—the Federal source for science about the Earth, its natural and living resources, natural hazards, and the environment, visit http://www.usgs.gov or call 1–888–ASK–USGS.

For an overview of USGS information products, including maps, imagery, and publications, visit http://www.usgs.gov/pubprod

To order this and other USGS information products, visit http://store.usgs.gov

Suggested citation:
Thelin, G.P., and Stone, W.W., 2013, Estimation of annual agricultural pesticide use for counties of the conterminous United States, 1992–2009: U.S. Geological Survey Scientific Investigations Report 2013-5009, 54 p.

FOREWORD

The U.S. Geological Survey (USGS) is committed to providing the Nation with reliable scientific information that helps to enhance and protect the overall quality of life and that facilitates effective management of water, biological, energy, and mineral resources (*http://www.usgs.gov/*). Information on the Nation's water resources is critical to ensuring long-term availability of water that is safe for drinking and recreation and is suitable for industry, irrigation, and fish and wildlife. Population growth and increasing demands for water make the availability of that water, measured in terms of quantity and quality, even more essential to the long-term sustainability of our communities and ecosystems.

The USGS implemented the National Water-Quality Assessment (NAWQA) Program in 1991 to support national, regional, State, and local information needs and decisions related to water-quality management and policy (*http://water.usgs.gov/nawqa*). The NAWQA Program is designed to answer: What is the quality of our Nation's streams and groundwater? How are conditions changing over time? How do natural features and human activities affect the quality of streams and groundwater, and where are those effects most pronounced? By combining information on water chemistry, physical characteristics, stream habitat, and aquatic life, the NAWQA Program aims to provide science-based insights for current and emerging water issues and priorities. From 1991 to 2001, the NAWQA Program completed interdisciplinary assessments and established a baseline understanding of water-quality conditions in 51 of the Nation's river basins and aquifers, referred to as Study Units (*http://water.usgs.gov/nawqa/studies/study_units.html*).

National and regional assessments are ongoing in the second decade (2001–2012) of the NAWQA Program as 42 of the 51 Study Units are selectively reassessed. These assessments extend the findings in the Study Units by determining water-quality status and trends at sites that have been consistently monitored for more than a decade, and filling critical gaps in characterizing the quality of surface water and groundwater. For example, increased emphasis has been placed on assessing the quality of source water and finished water associated with many of the Nation's largest community water systems. During the second decade, NAWQA is addressing five national priority topics that build an understanding of how natural features and human activities affect water quality, and establish links between sources of contaminants, the transport of those contaminants through the hydrologic system, and the potential effects of contaminants on humans and aquatic ecosystems. Included are studies on the fate of agricultural chemicals, effects of urbanization on stream ecosystems, bioaccumulation of mercury in stream ecosystems, effects of nutrient enrichment on aquatic ecosystems, and transport of contaminants to public-supply wells. In addition, national syntheses of information on pesticides, volatile organic compounds (VOCs), nutrients, trace elements, and aquatic ecology are continuing.

The USGS aims to disseminate credible, timely, and relevant science information to address practical and effective water-resource management and strategies that protect and restore water quality. We hope this NAWQA publication will provide you with insights and information to meet your needs, and will foster increased citizen awareness and involvement in the protection and restoration of our Nation's waters.

The USGS recognizes that a national assessment by a single program cannot address all water-resource issues of interest. External coordination at all levels is critical for cost-effective management, regulation, and conservation of our Nation's water resources. The NAWQA Program, therefore, depends on advice and information from other agencies—Federal, State, regional, interstate, Tribal, and local—as well as nongovernmental organizations, industry, academia, and other stakeholder groups. Your assistance and suggestions are greatly appreciated.

William H. Werkheiser
USGS Associate Director for Water

Contents

Contents—Continued

Figures

Figures—Continued

Tables

Conversion Factors

Inch/Pound to SI

Multiply	By	To obtain
Area		
acre	4,047	square meter (m^2)
acre	0.4047	hectare (ha)
acre	0.4047	square hectometer (hm^2)
acre	0.004047	square kilometer (km^2)
square mile (mi^2)	259.0	hectare (ha)
square mile (mi^2)	2.590	square kilometer (km^2)
Mass		
pound, avoirdupois (lb)	0.4536	kilogram (kg)

SI to Inch/Pound

Multiply	By	To obtain
Area		
square kilometer (km^2)	247.1	acre
square kilometer (km^2)	0.3861	square mile (mi^2)

Abbreviations

ACU — Agricultural Chemical Use
CDL — Cropland Data Layer
CRD — Crop Reporting District
DPR — California Department of Pesticide Regulation
DPR-PUR — Department of Pesticide Regulation-Pesticide Use Reporting (California)
EPest — Estimated pesticide use
FR — Fruitful Rim
GIS — Geographic Information System
NASS — National Agricultural Statistics Service
NAWQA — National Water Quality Assessment Program
NPUD — National Pesticide Use Data
PUR — Pesticide Use Reporting
RE — Relative error calculated as: (EPest–NASS)/NASS
USDA — U.S. Department of Agriculture
USEPA — U.S. Environmental Protection Agency
USGS — U.S. Geological Survey
WARP — Watershed Regressions for Pesticides

Estimation of Annual Agricultural Pesticide Use for Counties of the Conterminous United States, 1992–2009

By Gail P. Thelin and Wesley W. Stone

Abstract

A method was developed to calculate annual county-level pesticide use for selected herbicides, insecticides, and fungicides applied to agricultural crops grown in the conterminous United States from 1992 through 2009. Pesticide-use data compiled by proprietary surveys of farm operations located within Crop Reporting Districts were used in conjunction with annual harvested-crop acreage reported by the U.S. Department of Agriculture National Agricultural Statistics Service (NASS) to calculate use rates per harvested-crop acre, or an 'estimated pesticide use' (EPest) rate, for each crop by year. Pesticide-use data were not available for all Crop Reporting Districts and years. When data were unavailable for a Crop Reporting District in a particular year, EPest extrapolated rates were calculated from adjoining or nearby Crop Reporting Districts to ensure that pesticide use was estimated for all counties that reported harvested-crop acreage. EPest rates were applied to county harvested-crop acreage differently to obtain EPest-low and EPest-high estimates of pesticide-use for counties and states, with the exception of use estimates for California, which were taken from annual Department of Pesticide Regulation Pesticide Use Reports.

Annual EPest-low and EPest-high use totals were compared with other published pesticide-use reports for selected pesticides, crops, and years. EPest-low and EPest-high national totals for five of seven herbicides were in close agreement with U.S. Environmental Protection Agency and National Pesticide Use Data estimates, but greater than most NASS national totals. A second set of analyses compared EPest and NASS annual state totals and state-by-crop totals for selected crops. Overall, EPest and NASS use totals were not significantly different for the majority of crop-state-year combinations evaluated. Furthermore, comparisons of EPest and NASS use estimates for most pesticides had rank correlation coefficients greater than 0.75 and median relative errors of less than 15 percent. Of the 48 pesticide-by-crop combinations with 10 or more state-year combinations, 12 of the EPest-low and 17 of the EPest-high totals showed significant differences (p < 0.05) from NASS use estimates. The differences between EPest and NASS estimates did not follow consistent patterns related to particular crops, years, or states, and most correlation coefficients were greater than 0.75.

EPest values from this study are suitable for making national, regional, and watershed assessments of annual pesticide use from 1992 to 2009. Although estimates are provided by county to facilitate estimation of watershed pesticide use for a wide variety of watersheds, there is a greater degree of uncertainty in individual county-level estimates when compared to Crop Reporting District or state-level estimates because (1) EPest crop-use rates were developed on the basis of pesticide use on harvested acres in multi-county areas (Crop Reporting Districts) and then allocated to county harvested cropland; (2) pesticide-by-crop use rates were not available for all Crop Reporting Districts in the conterminous United States, and extrapolation methods were used to estimate pesticide use for some counties; and (3) it is possible that surveyed pesticide-by-crop use rates do not reflect all agricultural use on all crops grown. The methods developed in this study also are applicable to other agricultural pesticides and years.

Introduction

Hundreds of millions of pounds of pesticides are applied to agricultural crops every year to control weeds, insect infestations, plant diseases, and other pests. Annually, the total amount of conventional pesticides (excluding sulfur, petroleum oil, chlorine, hypochlorites, and wood preservatives) applied to crops grown throughout the conterminous United States has increased from a low of about 698 million pounds in the early 1990s (*http://www.epa. gov/opp00001/pestsales/07pestsales/historical_data2007_3. htm#table5_6*, accessed November 16, 2011) to a high of over 800 million pounds in 1996 (*fig. 1*). From 1996 through 2007, there was a slight downward trend in the total amount of pesticides used, reflecting decreases in the use of herbicides, plant growth regulators, and other conventional pesticides. Most of these differences in pesticide use can be attributed to changes in crop-management practices, the development of new pesticides that are effective at reduced use rates, and the introduction of genetically modified crops (Young, 2006; Fernandez-Cornejo and McBride, 2000).

Pesticides are important to crop management because they contribute to increased crop yields and improve the quality of crops. Pesticides applied to crops and soil, however, can be transported to surface water and groundwater, where they can degrade water quality. Pesticide concentrations in streams vary widely across the United States and are influenced by many factors, such as the amount and timing of pesticide applications and the soils, climate, and hydrology where they are applied (Gilliom and others, 2006). Nationally consistent information on the amount and geographic distribution of pesticide use, both current and historic, is essential for designing water-quality studies, interpreting water-quality data, assessing trends in pesticide use, and developing water-quality models that relate pesticide use to concentrations in the hydrologic environment.

Agricultural pesticide-use information is available from the U.S. Department of Agriculture (USDA) National Agricultural Statistics Service (NASS), but these data are reported as state totals for varying regions, crops, and years and, consequently, do not have sufficient geographic coverage, resolution, or temporal consistency to support studies at watershed or multicounty scales. California's Department of Pesticide Regulation (DPR) collects detailed pesticide-use information from all licensed applicators in the State and publishes annual Pesticide Use Reports (DPR-PURs)

that include detailed pesticide-use information (California Department of Pesticide Regulation, 2010). Agricultural pesticide-use data also are available from proprietary sources, but extrapolation techniques, such as those described in this report, are needed so that these data can be used by the National Water Quality Assessment (NAWQA) Program to estimate pesticide use for all counties of the conterminous United States.

A previous U.S. Geological Survey (USGS) study focused on developing extrapolation methods to determine county-level estimates for the herbicide atrazine by using proprietary pesticide-use reports and county harvested-crop acreage (Thelin and Stone, 2010). As part of that approach, regional rates were developed by using data from multiple years, and atrazine estimates were calculated for most counties in the conterminous Unites States. Comparisons with other data sources indicated that this approach to regional extrapolation could over-estimate pesticide use for pesticides that are not widely used across all geographic regions or when pesticide-use practices changed. This report describes an approach to estimating pesticide use, referred to as EPest, that is based on previous efforts but has changes that limit the use of regional rates, that incorporate a refined version of crop growing regions, and that expand the method to 39 herbicides, insecticides, and fungicides used in agriculture (*table 1*).

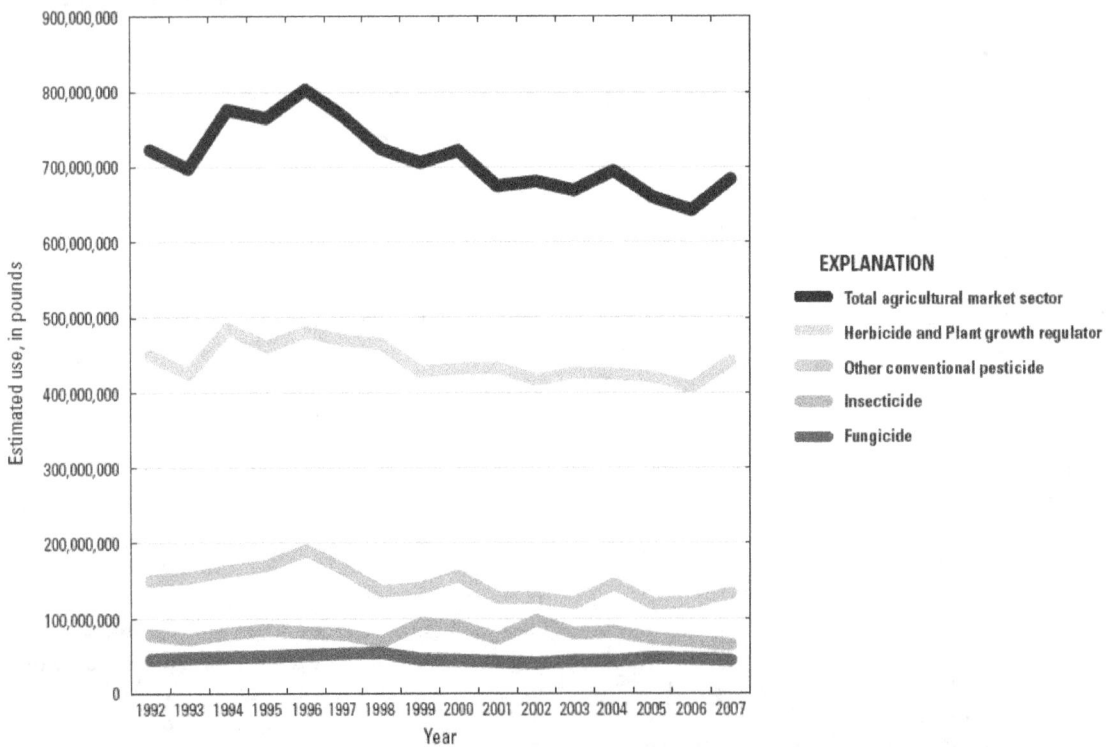

Figure 1.   Trends in agricultural conventional pesticide use in the conterminous United States, 1992–2009.

Table 1. List of pesticide names and type, for which annual county pesticide-use estimates were calculated.

Pesticide name	Type
Acetochlor	Herbicide
Acifluorfen	Herbicide
Alachlor	Herbicide
Atrazine	Herbicide
Benomyl	Fungicide
Bentazon	Herbicide
Bromoxynil	Herbicide
Butylate	Herbicide
Carbofuran	Insecticide
Chlorimuron	Herbicide
Cyanazine	Herbicide
EPTC	Herbicide
Ethalfluralin	Herbicide
Ethoprophos	Insecticide
Fluometuron	Herbicide
Fonofos	Insecticide
Glyphosate	Herbicide
Linuron	Herbicide
Methomyl	Insecticide
Methyl parathion	Insecticide
Metolachlor	Herbicide
S-metolachlor	Herbicide
Metribuzin	Herbicide
Nicosulfuron	Herbicide
Norflurazon	Herbicide
Oryzalin	Herbicide
Oxamyl	Insecticide
Pebulate	Herbicide
Phorate	Insecticide
Propachlor	Herbicide
Propanil	Herbicide
Propargite	Insecticide
Propiconazole	Fungicide
Propyzamide	Herbicide
Terbacil	Herbicide
Terbufos	Insecticide
Thiobencarb	Herbicide
Triallate	Herbicide
Trifluralin	Herbicide

Purpose and Scope

The purpose of this report is to describe (1) a method to estimate annual pesticide-by-crop use rates (pounds applied per harvested-crop acre), referred to as EPest rates, for 39 pesticides; (2) the process that was followed to apply these rates to produce an EPest-low and EPest-high estimate of annual use for each county; and (3) how the estimates for selected pesticides and crops derived by these methods compare with estimates from other published sources. This method was developed by using pesticide-use estimates reported for Crop Reporting Districts (CRDs) to calculate annual pesticide-by-crop use rates and, from that, estimates of pesticide use for individual counties. The 39 selected pesticides represent some of the primary pesticides used throughout the nation on row crops and several orchard and

vegetable crops, and include 28 herbicides, 9 insecticides, and 2 fungicides. Most of these same pesticides were included in a Watershed Regressions for Pesticides (WARP) multi-compound model analysis (Charles Crawford, U.S. Geological Survey, oral commun., 2011).

The pesticides evaluated in this study represent a range of herbicides, insecticides, and fungicides that are used on a variety of row, fruit, nut, and specialty crops grown in different environmental settings. Several of these pesticides have had changes in use over time, providing an evaluation of method performance for a wide range of conditions. To assess the accuracy of EPest totals, state-level totals were compared with NASS use estimates for selected pesticides and crops for states and years for which NASS survey data were available.

Data Sources

Data sources used to develop EPest pesticide-by-crop use rates and annual pesticide-use estimates by county included the following: (1) proprietary pesticide-by-crop use estimates reported for CRDs; (2) USDA county harvested-crop acreage reported in the 1992, 1997, 2002, and 2007 Census of Agriculture (*http://www.agcensus.usda.gov/*), and NASS annual harvested-crop acreage data collected from crop surveys for non-census years (*http://quickstats.nass.usda. gov/*); (3) boundaries for CRDs and counties; (4) regional boundaries derived from USDA Farm Resource Regions; and (5) pesticide-use information from California DPR-PUR. Each of these sources is described in following sections.

Pesticide-Use Data

Proprietary data from GfK Kynetec, Inc. on the amounts of pesticides applied to individual crops by CRD are the primary source of information used in this study and are referred to as surveyed use data in the remainder of this report. The surveyed use data are based on agricultural pesticide use surveys of more than 20,000 farm operations distributed throughout the conterminous United States (AgroTrak Quality Management Plan, written commun., August 2011). Data from the Census of Agriculture on the size (in acres) and number of farms that grow individual crops and represent selected land uses, such as pasture, are used to stratify all farms in the United States by size and to allocate the number of farms that will be surveyed in each strata. The survey design allocates a greater proportion of the sample to larger farm operations so that a greater percentage of crop acreage is represented, with the goal of more accurate characterization of farm operations and pesticide-use patterns. Use estimates for over 400 pesticides that are applied to a variety of row, specialty, fruit, and nut crops are reported by multi-county areas, referred to as CRDs (*fig. 2*). Surveys of farm operations within each CRD are extrapolated to represent total pesticide use for that CRD, and then estimates for individual CRDs or groups of CRDs are expanded to estimate pesticide use for states.

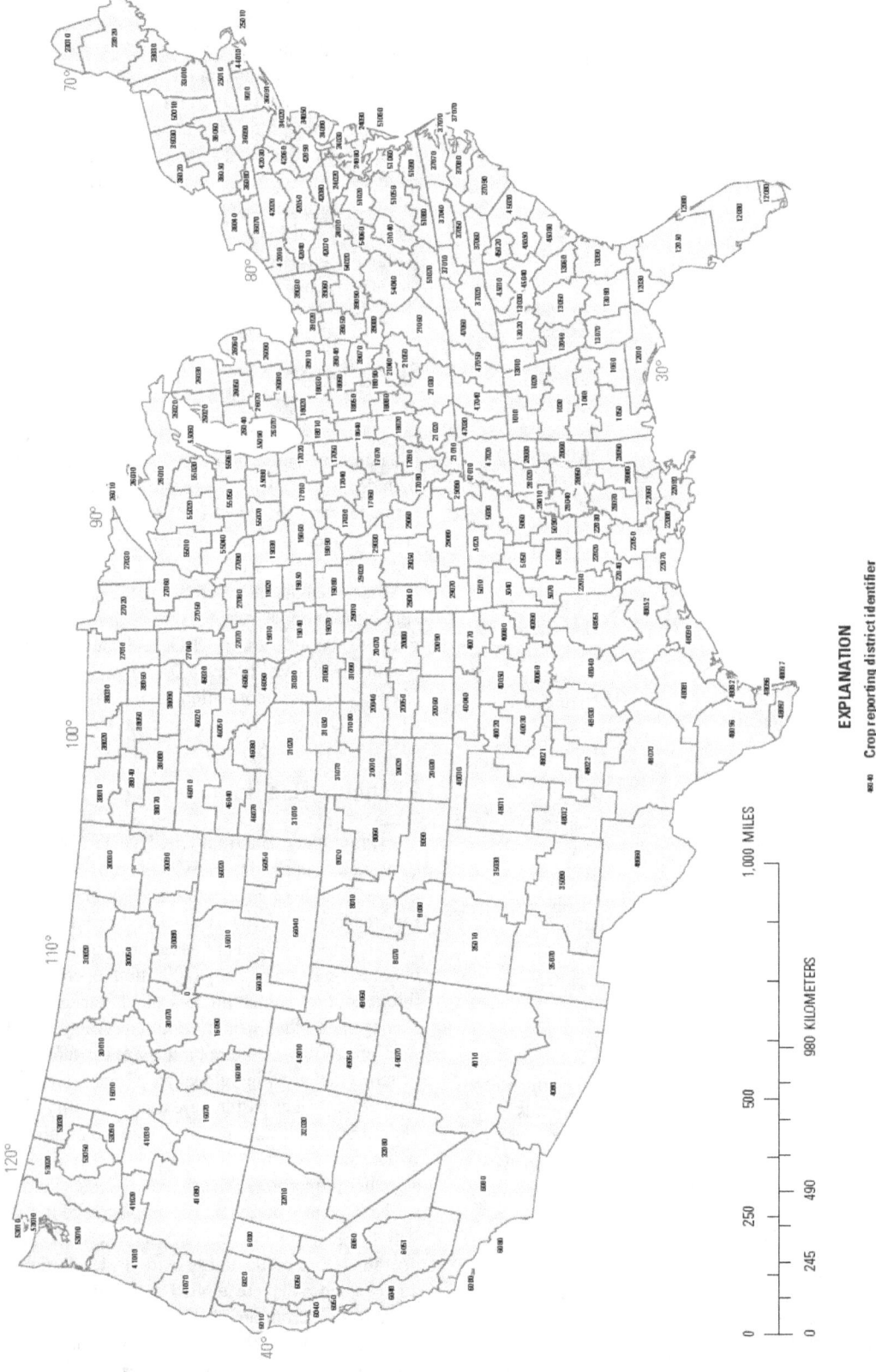

EXPLANATION

48|40 • Crop reporting district identifier

Figure 2. Crop Reporting Districts of the conterminous United States.

Harvested-Crop Acreage

The surveyed use data are based on planted-crop acres within a CRD, but NAWQA requires pesticide-use estimates at the county scale, including use estimates for pesticides that potentially were not surveyed. Therefore, the surveyed use data had to be disaggregated from CRDs to the individual counties. The USDA is the only uniform source of annual crop-acreage estimates for all counties in the United States. The USDA reports data on planted and harvested-crop acreage, but planted-acreage data are not available from the USDA for all of the individual crops with surveyed use data. Therefore, harvested acreage, rather than planted acreage, was used to develop annual pesticide-by-crop use rates. In taking this approach, it is recognized that use-rate estimates could be numerically greater than actual use rates on planted crops because not all planted acres are harvested. The emphasis of the method was to develop the best possible estimates of total use in a county, which required the use of the comprehensive data on harvested cropland. Annual harvested-crop acreage by county data from the USDA Census of Agriculture and NASS crop surveys were used in method development (1) to calculate the pesticide-by-crop use rates for each crop and CRD surveyed, and (2) to estimate pesticide use for all counties that report harvested acreage in the conterminous United States. Harvested-crop acreage was obtained from the Census of Agriculture for 1992, 1997, 2002, and 2007, and from NASS annual surveys for the years between censuses. *Table 2* lists the crops for which EPest use rates were developed and the USDA crop names for which acreage data were retrieved from the Census of Agriculture and NASS.

County-level harvested-crop acreage for the 76 crops and other non-crop agricultural-land uses, such as pasture and woodland, were obtained from USDA reports and used to produce harvested-crop acreage totals for all CRDs. However, additional processing was required in three cases: (1) the USDA did not report county acreage for a crop and year because of census nondisclosure rules that protect the identity of individual farm operations, (2) the USDA-NASS annual surveys did not collect data for a particular state or crop, or (3) the crop acreage was the total acreage for multiple categories of that crop. In cases when county acreage was not reported because of USDA nondisclosure rules or when a crop and state had not been surveyed by NASS, the county crop acreage was estimated through linear interpolation of acreage reports for the crop and county from consecutive years before and after the year of missing crop acreage. In order to produce acreage totals for EPest crop names that were composed of more than one USDA crop name, the subcategories for that crop were summed to produce total harvested acreage. For example, the county total for sorghum acreage was calculated by summing the acreage for the subcategories of sorghum:

sorghum for grain, sorghum for silage, and sorghum for syrup. Crop-acreage totals that comprised more than one crop name typically required crop acreage to be estimated through linear interpolation for some of the crop names because NASS crop surveys do not report all the same crop names as the Census of Agriculture. For example, NASS did not report acreage of corn for forage from 1992 through 2009. To estimate corn-for-forage acreage in non-census years, the acreage from two Censuses of Agriculture (prior and next) was interpolated to fill in the non-surveyed corn-for-forage acreage.

Geospatial Data

Two geospatial datasets were integral to the method used to calculate pesticide-by-crop use rates for surveyed and non-surveyed CRDs. These datasets included boundaries for CRDs and USDA Farm Resource Regions (*http://www.ers. usda.gov/Briefing/ARMS/resourceregions/resourceregions. htm*). CRD boundaries were used (1) to develop a table that listed the spatial relation of each CRD in the conterminous United States to its surrounding CRDs and (2) to determine the counties that were associated with each CRD so that estimates reported for CRDs could be disaggregated to counties. The second geospatial dataset was a modified version of the USDA Farm Resource Boundaries, which was used (1) to determine the Farm Resource Region for each CRD and (2) to develop regional use rates for individual crops when a CRD rate did not exist.

CRDs are defined as multi-county areas that share similar geographic attributes, including soil type, terrain, elevation, and climatic factors, such as mean temperature, annual precipitation, and length of growing season. There are 304 CRDs in the conterminous United States, and most states are divided into 9 CRDs; however, some states, such as Massachusetts and New Hampshire, contain only 1 CRD, whereas Texas has 15 CRDs.

A geospatial vector dataset of CRD boundaries was used to generate a table that enumerates the spatial relation between each of the individual CRDs and the CRDs surrounding each of these 'primary' CRDs. For each primary CRD, two concentric rings of CRDs were identified by using a Geographic Information System (GIS) proximity mapping function. CRDs that touched the primary CRD were designated as tier 1 CRDs, and CRDs that touched tier 1 CRDs were designated as tier 2 CRDs. Any CRD could be considered a primary, a tier 1, or a tier 2 CRD, depending on which CRD is central to the area of interest. *Figure 3*, for example, shows primary CRD 20060 (Kansas CRD 60) and the tier 1 and tier 2 CRDs that are associated with it. When CRD-level pesticide use data were not available, associated tier 1 and tier 2 CRDs were used to calculate pesticide-by-crop rates.

Table 2. EPest crop name and corresponding U.S. Department of Agriculture (USDA) Census of Agriculture crop names.

EPest crop name	USDA, Census of Agriculture crop name(s)
Alfalfa	Alfalfa hay
Almonds	Almonds
Apples	Apples
Barley	Barley for grain
Beans and peas	Green lima beans; snap beans; green peas, excluding southern peas; peas, green southern
Berries	Strawberries
Bulb crops	Garlic; green onions; dry onions
Conservation Reserve Program (CRP), long-term acres	Land enrolled in conservation reserve or wetlands reserve programs
Canola, rapeseed	Canola, other rapeseed
Cherries	Sweet cherries; tart cherries
Citrus, other	Other citrus fruit
Cole crops	Broccoli
Corn	Corn for grain
Cotton	Cotton, all
Cropland for pasture	Cropland used for pasture or grazing
Cucurbits	Cucumbers and pickles; pumpkins; squash
Dry beans and peas	Dry lima beans; dry edible beans, excluding limas; dry edible peas; dry southern peas
Eggplant and peppers	Eggplant; peppers, bell; peppers, chile
Summer fallow	Summer fallow
Flax	Flaxseed
Grapefruit	Grapefruit
Grapevines	Grapes
Hay, other	Grass silage, haylage
Idle cropland, other	Idle cropland, other
Leafy vegetables, other	Celery; spinach
Lemons	Lemons
Lettuce	Lettuce all
Lots, farmsteads, other	Lots, farmsteads and other
Melons	Cantaloupes; watermelons
Nut trees, other	Hazel nuts (filberts); pistachios
Oats and rye	Oats for grain; rye for grain
Oranges	Oranges, all
Pasture/range	Pastureland and rangeland, other than cropland and woodland pastured
Peaches	Peaches, all
Peanuts	Peanuts for nuts
Pears	Pears, all
Pecans	Pecans
Potatoes	Potatoes
Prunes	Plums and prunes
Rice	Rice
Roots and tubers	Carrots
Sorghum	Sorghum for grain; sorghum for sileage or green chop; sorghum for syrup
Soybeans	Soybeans for beans
Stone-like fruit, other	Apricots; avocados
Sugarbeets	Sugar beets for sugar
Sugarcane	Sugar cane for sugar
Sunflowers	Sunflower seed all
Sweet corn	Sweet corn
Tobacco	Tobacco
Tomatoes	Tomatoes
Other vegetables	Artichokes
Walnuts	Walnuts, english
Wheat, spring	Durum wheat for grain; other spring wheat for grain
Wheat, winter	Winter wheat for grain
Woodland	Total woodland

A geospatial dataset of USDA Farm Resource Regions was used to develop regional pesticide-by-crop use rates for CRDs that were not surveyed and for which a tier 1 or tier 2 rate was not available. In a previous atrazine study (Thelin and Stone, 2010), USDA Farm Production Regions were used to develop regional rates. These boundaries follow state boundaries and often combine large areas that can have different soils, topography, and agricultural practices. The Farm Production Region boundaries were replaced with USDA Farm Resource Regions because these boundaries take into account farm practices and physiographic, soil, and climatic traits (*http://www.ers.usda.gov/publications/aib760/aib-760.pdf*). Farm Resource Region boundaries conform to CRD boundaries. There are nine Farm Resource Regions, which were further subdivided in cases where the region was not contiguous. For example, the Fruitful Rim (FR) Region is located in parts of the West, Southwest, and Southeastern United States, so this large region was subdivided into four subregions: (1) FR-Northwest, including Washington and parts of Oregon and Idaho; (2) FR-West, including parts of California and Arizona; (3) FR-Texas, including Texas and

New Mexico; and (4) FR-Southeast, including Florida and parts of Alabama, Georgia, and South Carolina. Similarly, the Eastern Uplands, Northern Crescent, and Southern Seaboard were divided into eastern and western subregions (*fig. 4*).

Pesticide-Use Estimates for California

EPest-low and EPest-high estimates for California were not calculated by using the method described in this report; instead, county totals were obtained from the online DPR-PUR database (California Department of Pesticide Regulation, 2010). Since 1990, California has required reporting of all agricultural pesticide use. DPR-PUR includes information on the pesticide applied, location and time of application, and the agricultural crop treated. Annual pesticide-use estimates by crop were retrieved from the online DPR-PUR database and merged with the EPest-low and EPest-high county data after the estimation process was completed for the rest of the country.

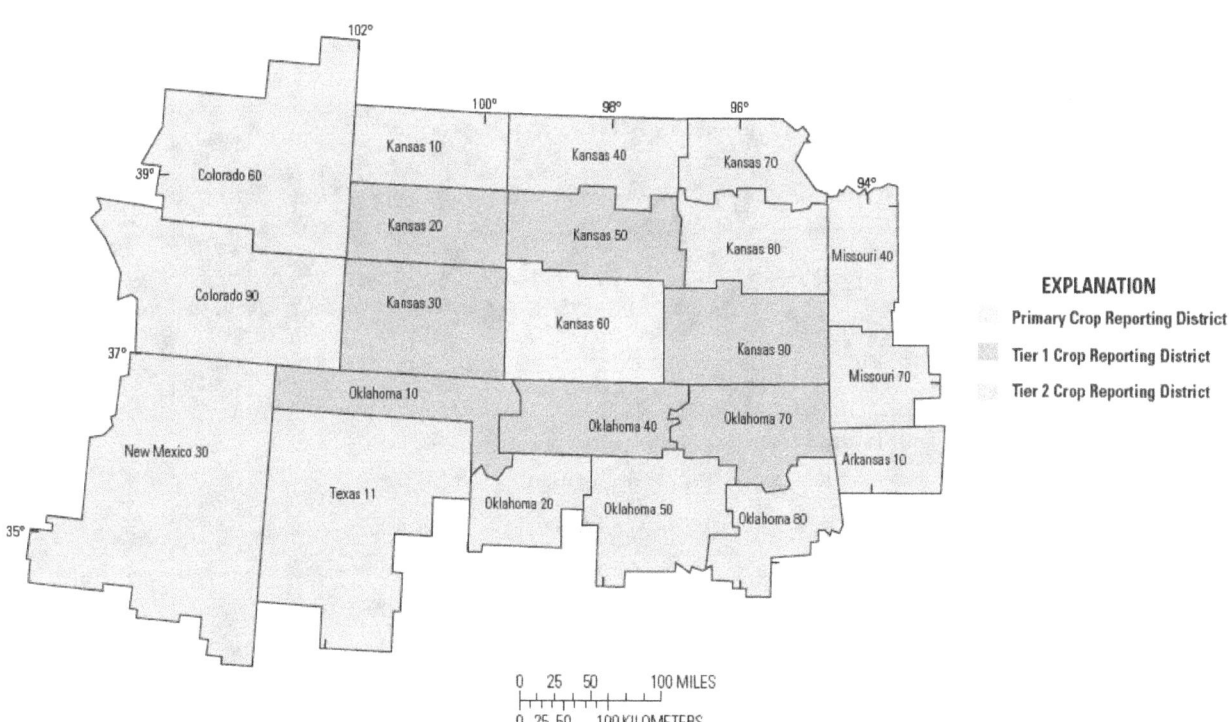

Figure 3. Crop Reporting District 20060 (Kansas CRD 60) and neighboring tier 1 and tier 2 Crop Reporting Districts.

Methods for Estimating Pesticide Use

The following sections describe methods developed to estimate agricultural pesticide use for counties in the conterminous United States, except those in California. In order to calculate estimates of pesticide use for counties, pesticide-by-crop use rates were developed for CRDs on the basis of surveyed use data and harvested-crop acreage from the USDA. The resulting pesticide-by-crop use rates are referred to as EPest surveyed-use rates, which are calculated by dividing the amount of pesticide applied to a crop in the CRD by harvested-crop acres. Not every CRD in the conterminous United Sates was surveyed; therefore, EPest

extrapolated rates were developed for unsurveyed CRDs by using surveyed rates from nearby CRDs or surveyed and extrapolated rates from CRDs in the same region. A surveyed or an extrapolated rate, depending on the CRD, was applied to county harvested acreage to estimate pesticide use on individual crops grown in each county of the conterminous United States, except California. The following sections describe (1) the method used to replace false zero values reported in the surveyed use data with inferred data, (2) how the EPest surveyed and extrapolated rates were developed, and (3) the decision process that was followed to assign these EPest rates to counties to produce EPest-high and EPest-low estimates of pesticide use for counties in the conterminous United States.

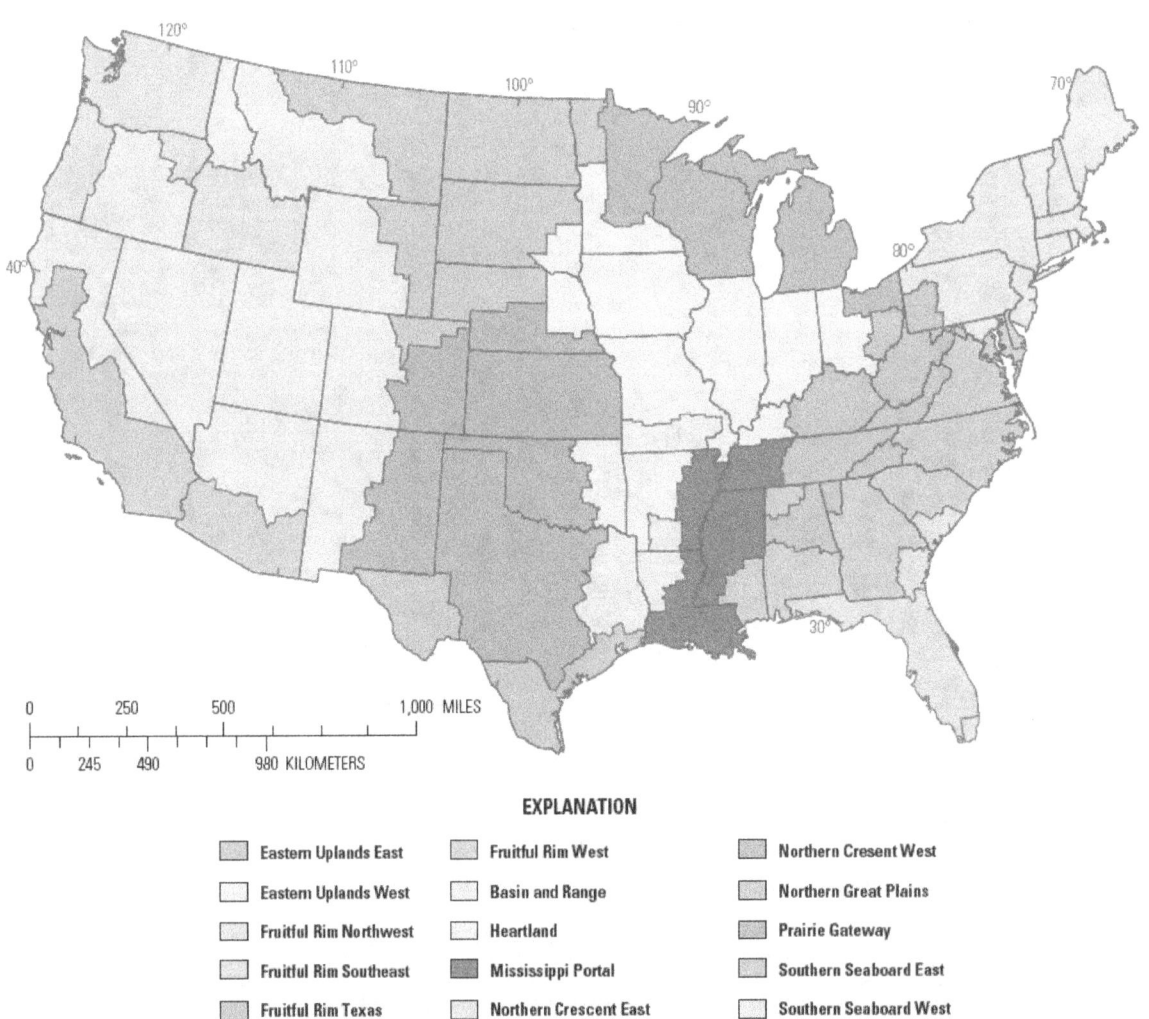

Figure 4.   U.S. Department of Agriculture Farm Resources Regions (*http://www.ers.usda.gov/publications/aib760/aib-760.pdf*), as subdivided for calculating regional estimated pesticide-use rates.

Processing Zero Values

The surveyed-use data included the following elements: pounds of pesticide applied to a crop, number of crop acres treated, and overall pesticide-by-crop application rate. In some cases, a zero value was reported for one or more of the data elements because of rounding or truncating values of less than one; therefore, a new inferred value was calculated to replace the false zero values as follows:

1. When the pounds applied were reported as zero, but the number of acres treated was greater than zero, and an application rate was reported, then a value for the pounds applied was calculated by multiplying the number of acres treated by the pesticide-by-crop application rate reported for the surveyed CRD.

2. When the number of acres treated and the pounds applied were reported as zero for the surveyed CRD, but an application rate was reported, then it was assumed that the number of acres treated was equal to one, and the pounds applied were equal to the application rate for 1 acre as reported for the CRD.

3. When the pounds applied and application rate were reported as zero for the surveyed CRD, but the number of acres treated was greater than zero, a new application rate could not be calculated. In these cases, the lowest non-zero application rate in the surveyed-use data across all years, pesticides, crops, and CRDs, which was 0.001 pounds per acre annually, was used to estimate the pounds applied (0.001 pounds per acre multiplied by the number of acres treated).

EPest Crop-Use Rates for Surveyed CRDs

EPest surveyed rates for 1992 through 2009 were developed for each of the 39 pesticides included in this study by using surveyed-use estimates of pounds of pesticides applied to individual crops and the harvested acreage for these crops reported by USDA. The pesticide-by-crop use rates determined from surveyed-use data for CRDs are based on planted-crop acreage, but were adjusted to harvested acreage for EPest county-level pesticide-by-crop use rates. EPest surveyed pesticide-by-crop use rates were calculated by dividing the pounds of pesticide applied to a crop in a CRD by the harvested-crop acreage in the CRD to yield a use rate per harvested acre—for a specific crop this is referred to as an EPest surveyed pesticide-by-crop use rate. Use rates calculated by using harvested-crop acreage rather than planted acreage can result in a greater rate per acre because, typically, there are fewer harvested acres than planted acres as a result of crop failure. To avoid artificially high use rates caused solely by the difference between planted and harvested acres, the harvested-crop acreage for the CRD and associated counties was adjusted if the CRD harvested-crop acres were less than

the surveyed CRD planted-crop acres. Specifically, a county-CRD weighting factor for each crop and year was calculated by determining the percentage that each county's acreage contributed to the total acreage in the CRD. When the sum of the harvested-crop acreage for counties in the CRD was less than the planted-crop acreage for the CRD reported in the surveyed-use data, the weighting factor was used to adjust the harvested acreage for each county in the CRD to the survey-reported planted-crop acreage.

EPest Use Rates for Unsurveyed CRDs—Tier 1, Tier 2, and Regional Use Rates

EPest surveyed-use rates were applied to the harvested-crop acreage in all counties that were part of the surveyed CRDs. Some CRDs, however, were not surveyed for a particular year or combination of years, even though a pesticide could have been used there. For these CRDs, indirect estimates were derived. To ensure that pesticide-use estimates accounted for all acreage that could have been treated, extrapolated use rates were developed for individual pesticides and crops in unsurveyed CRDs through a set of decision rules (*fig. 5*).

The decision process included developing three types of extrapolated pesticide-by-crop use rates, referred to as tier 1, tier 2, and regional rates. How a use rate was estimated for an unsurveyed CRD depended on the availability of rates from surrounding tier 1 and tier 2 CRDs. For this purpose, the proximity table of CRDs, described previously, was searched to determine if a new rate could be calculated on the basis of rates from tier 1 or tier 2 CRDs. First, the tier 1 CRDs surrounding the unsurveyed CRD were searched, and if one or more surveyed pesticide-by-crop use rates existed, the median rate was used from these surveyed rates, called tier 1 EPest rate, to estimate pesticide-by-crop use for the counties in the unsurveyed CRD. If a tier 1 rate could not be established because there were no surveyed rates available, then tier 2 CRDs were searched to determine if three or more of the tier 2 CRDs had surveyed rates. If so, then the median value of these rates was used as the tier 2 EPest rate which was then applied to the counties in the unsurveyed CRD. Finally, if a tier 1 or tier 2 EPest rate could not be determined, then a regional rate was calculated for the modified USDA Farm Resource Region (described previously) and used for the CRD. Regional rates were the median of all non-zero EPest rates, including surveyed, tier 1, and tier 2 EPest from the same modified USDA Farm Resource Region. To reduce the influence of duplicate extrapolated EPest rates on the calculation of regional rates, duplicate extrapolated rates were removed prior to the calculation. *Figure 6* illustrates the process of establishing and assigning EPest extrapolated rates for counties in the Southern Seaboard Region-East by using *S*-metolachlor on corn as an example.

The Southern Seaboard-East region is composed of 36 CRDs from all or part of 8 states, including Alabama, Delaware, Georgia, Maryland, Mississippi, North Carolina, South Carolina, and Virginia (fig. 6). In 2007, there were surveyed-use data for S-metolachlor on corn in 17 of the 36 CRDs in the region. On the basis of the surveyed rates for the 17 surveyed CRDs, S-metolachlor use on corn was estimated for 180 of 388 counties in the region. There were an additional 208 counties in the region that had corn acreage, but a surveyed rate was not available, so EPest tier 1, tier 2, or regional rates were estimated as described in the following paragraphs.

Tier 1 S-metolachlor-corn rates were estimated for 11 CRDs in the example region and applied to 114 counties in these CRDs. South Carolina CRD 45030, labeled A in figure 6, is used to illustrate how a tier 1 rate is calculated from adjacent tier 1 CRDs. The tier 1 rate was developed for South Carolina CRD 45030 by using surveyed rates from three surrounding CRDs, which had EPest surveyed rates of 0.0095, 0.7093, and 1.123 pounds per harvested acre. There were two other CRDs adjacent to South Carolina 45030, but there were no surveyed rates available for them. In this example, the median of the three available EPest surveyed rates was 0.7093 pounds per harvested acre (North Carolina CRD 37090), and this rate was used as the tier 1 rate to estimate 2007 S-metolachlor use on corn in the nine counties that are part of South Carolina CRD 45030.

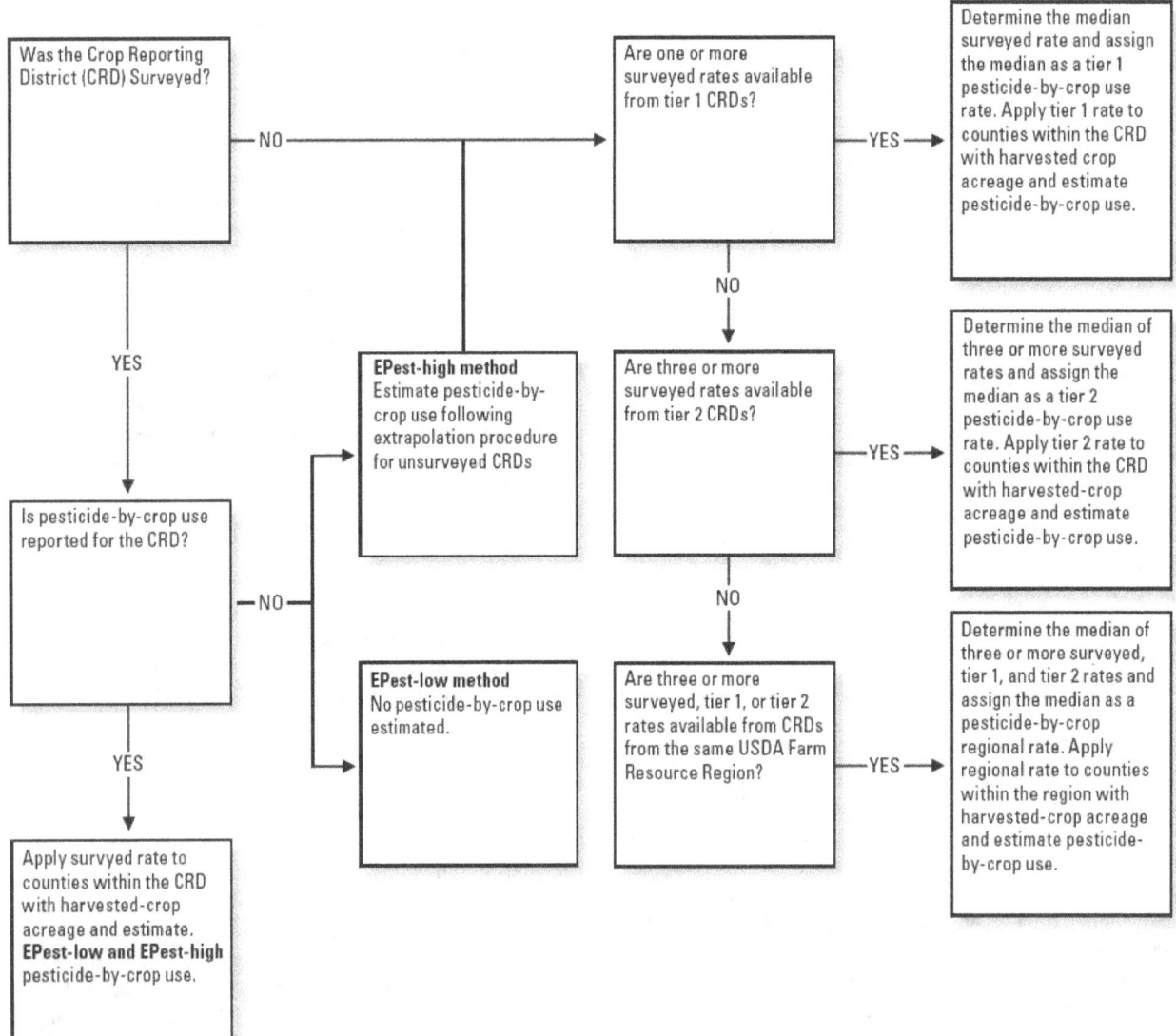

Figure 5. Summary of decision process followed to develop EPest rates.

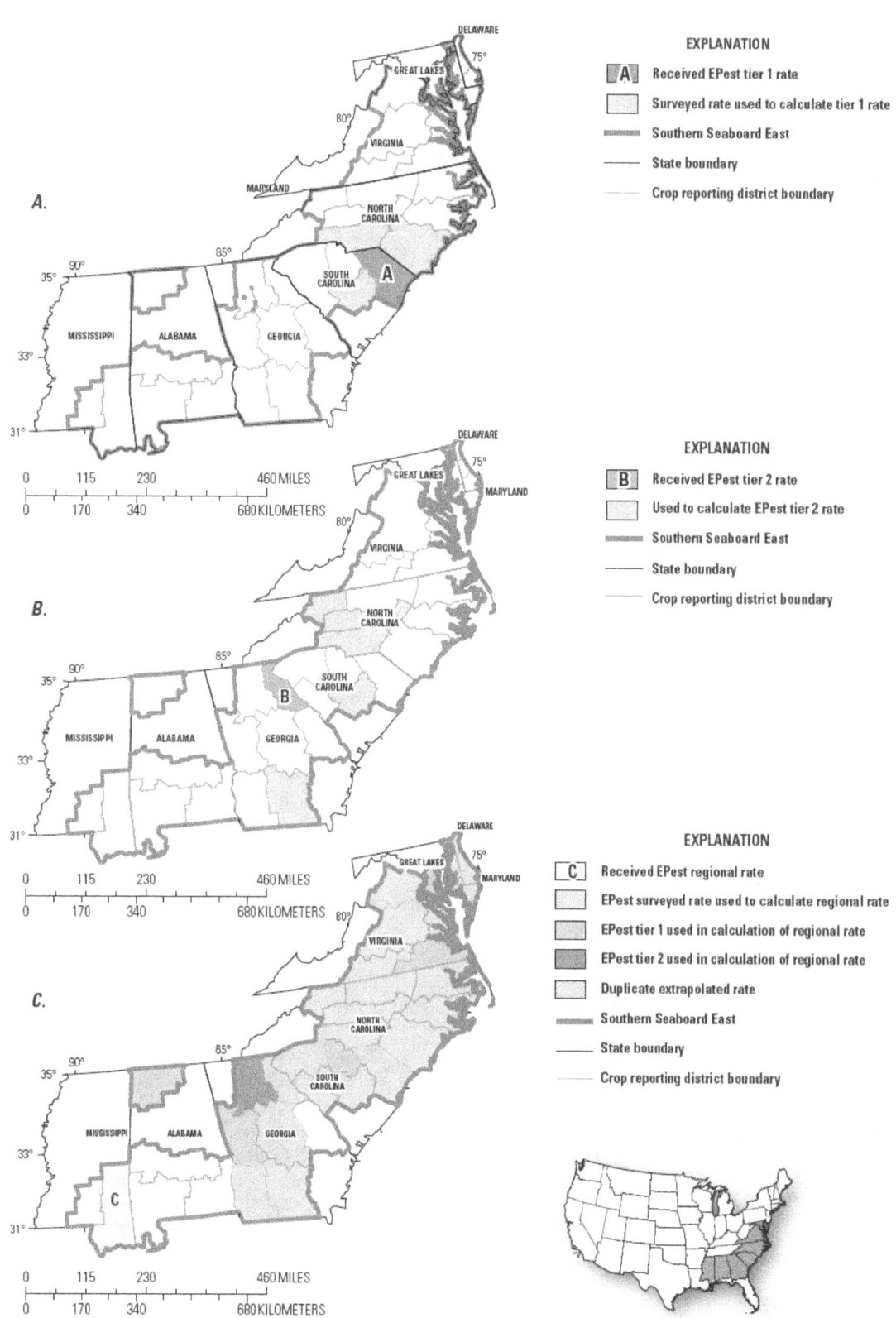

Figure 6. Methods for establishing extrapolated estimates for 2007 *S*-metolachlor use on corn in the Southern Seaboard-East region for (*A*) EPest tier 1 rate, (*B*) EPest tier 2 rate, and (*C*) EPest regional rate.

In the Southern Seaboard-East region, tier 2 *S*-metolachlor rates for corn were applied to 25 counties in two CRDs. Georgia CRD 13030, labeled *B* in *figure 6*, is an example of determining a tier 2 rate from surrounding CRDs. There were no EPest surveyed rates for *S*-metolachlor-corn from adjacent CRDs, so tier 2 CRDs were used. A minimum of three rates are required to determine a tier 2 rate, and there were five tier 2 CRDs that had surveyed annual rates of 0.1445, 0.3156, 0.7009, 0.9565, and 1.1229 pounds per harvested acre. The median of these five rates was 0.7009 pounds per harvested acre, which was assigned as the tier 2 rate used to estimate 2007 *S*-metolachlor use on corn for the nine counties in Georgia CRD 13030.

Finally, regional rates were calculated for 2007 *S*-metolachlor-corn in the Southern Seaboard-East region and applied to 6 CRDs and 69 counties. Mississippi CRD 28009, labeled *C* in *figure 6*, is used to illustrate how the regional rate was calculated from adjacent surveyed, tier 1, and tier 2 CRDs. There were 30 EPest rates available for the region, including 17 surveyed rates, 11 tier 1 rates, and 2 tier 2 rates. In the calculation of a regional rate, a minimum of three surveyed, tier 1, or tier 2 rates are required, and any duplicate extrapolated rates are dropped prior to calculating the median. In calculating the median regional rate, 7 duplicate rates were dropped, including 6 tier 1 rates and 1 tier 2 rate, so that 17 surveyed rates, 5 tier 1 rates, and 1 tier 2 rate were used to find the 2007 median rate of 0.3069 pounds per harvested acre of corn.

EPest-Low and EPest-High Estimates

Two variations on the method for estimating county pesticide use were developed to yield EPest-low and EPest-high estimates for counties in the conterminous United States other than California. Both methods incorporated surveyed and extrapolated rates to estimate pesticide use for counties, but EPest-low and EPest-high estimations differed in how they treated situations when a CRD was surveyed and pesticide use was not reported for a particular pesticide-by-crop combination (*fig. 5*). If use of a pesticide on a crop was not reported in a surveyed CRD, EPest-low reports zero use in the CRD for that pesticide-by-crop combination. EPest-high, however, treats the unreported use for that pesticide-by-crop combination in the CRD as unsurveyed, and pesticide-by-crop use rates from neighboring CRDs and, in some cases, CRDs within the same USDA Farm Resource Region are used to calculate the pesticide-by-crop EPest-high rate for the CRD.

Results

EPest-low and EPest-high totals were calculated from 1992 through 2009 for the 39 selected pesticides by using the methods described in this report. EPest-low totals, including California, were available for a low of 3,021 counties in 2008 to a high of 3,056 counties in 1992. The EPest-high method produced estimates for 3,049 counties in 2000 and 3,060 counties in 1994, including those in California. Pesticide-use estimates for counties in California are available from 1992 through 2009 for 35 of the 39 pesticides in this study. Use estimates are not available for the pesticides acetochlor, chlorimuron, propachlor, and terbufos because these pesticides were not used in California. For counties in California, there is a single county estimate, rather than a high and low estimate per pesticide by crop and year, which represents the sum of individual pesticide applications in a county reported by DPR-PUR (*ftp://pestreg.cdpr.ca.gov/pub/ outgoing/pur_archives*).

EPest-low and EPest-high county pesticide-use totals for 1992–2009 are available from *http://water.usgs.gov/ nawqa/pnsp/usage/maps/*. The county estimates represent the sum of individual pesticides used on all row, fruit, nut, and vegetable crops and selected agricultural land uses, such as summer fallow, pasture, and woodland. *Appendix 1* provides the annual EPest-low and EPest-high national totals for each of the 39 pesticides, the total pounds applied to individual crops, and the percentage of the national pesticide total each crop represents. With the exception of acetochlor, fonofos, propachlor, and *S*-metolachlor, annual estimates are available for 1992 through 2009. Acetochlor estimates are available beginning in 1994, when it was first registered for use, while estimates for fonofos and propachlor are reported for 1992 through 2005, and *S*-metolachlor estimates are available beginning in 1997.

EPest-low and EPest-high national use totals for each of the 39 pesticides are shown in *appendix 2* along with the amount and percentage of the total estimate that was derived from EPest surveyed, tier 1, tier 2, and regional rates, and from the DPR-PUR for California. Across all pesticides and years, the amount added to the EPest-low national total by extrapolated tier 1, tier 2, or regional rates, ranged from less than 1 percent for most compounds for one or more years to as much as 36 percent for terbacil use in 2003. A greater proportion of the EPest-high national total was derived from extrapolated rates, which ranged from less than 1 percent to as much as 94 percent for butylate use in 2007.

About 23 percent of the EPest-low and EPest-high annual national use totals were within 10 percent of one another and about 45 percent were within 25 percent of one another. EPest-high totals were more than double EPest-low totals for the pesticides alachlor, butylate, carbofuran, cyanazine, ethoprophos, linuron, methyl parathion, metolachlor, pebulate, propachlor, and terbacil for at least six of the years estimated. The extrapolated rates for surveyed CRDs used in EPest-high methods more than doubled the national total pesticide use for some years and pesticides for some specialty crops; for major crops, such as corn and alfalfa; and for some land uses, such as summer fallow, pasture and rangeland.

For the pesticides included in this study, EPest-low annual-use totals were less than or equal to EPest-high annual-use totals, as shown in *appendix 2*. However, EPest-low annual-use totals can be greater than EPest-high totals when the EPest-low pesticide-by-crop regional rate is greater than the EPest-high rate. EPest regional pesticide-by-crop rates are determined by using a minimum of three CRDs, and, typically, EPest-high regional rates were determined from a greater number of CRDs than EPest-low regional rates. In some cases, rates from additional CRDs can result in an EPest-high regional pesticide-by-crop rate that is less than the EPest-low regional rate. For example, if the EPest-low regional rate were determined from five rates—158, 54, 31.8, 9.68, and 5 pounds per acre—then the median would be 31.8 pounds of pesticide per harvested acre. The rates from these same five CRDs along with the EPest-high rates from any other CRDs in the region would be used to calculate the EPest-high regional rate. For example, if 158, 54, 31.8, 9.68, 9.05, 6.7, and 5 pounds of pesticide per crop acre were the rates used to determine the EPest-high regional rate, the EPest-high pesticide-by-crop regional rate would be 9.68 pounds of pesticide per harvested acre. Although these two rates were for the same counties in the region, the EPest-low total would be greater than the EPest-high use total.

In cases when a CRD was not surveyed, and a tier 1, tier 2, or regional rate was available, both EPest-low and EPest-high methods determined a pesticide-by-crop rate. In general, extrapolated rates for non-surveyed CRDs represented a greater percentage of use in more recent years because some pesticides were reported less frequently and some crops were not surveyed as extensively. EPest tier 1, tier 2, and regional rates have inherently greater uncertainty than rates for surveyed CRDs because a pesticide could have been applied to a localized area in response to a pest infestation, while the same crop grown in another part of the same region would not be managed in the same way, which can result in misrepresentative estimates of pesticide use. In addition, some EPest-high annual totals for pesticides that have been replaced or phased out, such as metolachlor and cyanzine, can be inaccurate because the EPest-high method assumes if a CRD was surveyed and an estimate for the pesticide was not reported, then an extrapolated rate could be used to estimate pesticide use.

Comparison of EPest National Estimates with Other Sources

National annual pesticide-use estimates developed by using EPest-low and EPest-high methods were compared with independently published estimates for seven herbicides. These comparisons were limited to acetochlor, alachlor, atrazine, EPTC, glyphosate, propanil, and trifluralin and to selected years because of limited data from the published sources. EPest totals for 1997, 2001, and 2007 were compared to (1) agricultural-use estimates published by the U.S. Environmental Protection Agency (USEPA; Kiely and others, 2004; Grube and others, 2011), (2) NASS-Agricultural Chemical Use (ACU) data (National Agricultural Statistics Service, 2008; hereinafter, referred to as NASS), and (3) National Pesticide Use Database (NPUD) estimates (Crop Protection Research Institute, 2006). NASS annual data were published as the "Total of Program States" in pounds per year and represent the amount of pesticide estimated for the states and crops that were surveyed for a specific year. Thus, the NASS national totals shown in these analyses are not intended to represent total use for all states or crops but are included as a point of reference. The USEPA estimates were reported as a range for each pesticide on agricultural crops as determined from a variety of public and proprietary data sources. Estimates for some pesticides and years were not available for each set of analyses, so comparisons were made for the years with the most complete data from each of the sources. Annual state estimates for the pesticides compared were available from EPest for 1992 through 2009; USEPA for 1997, 2001, 2003, 2005, and 2007; NPUD for 1992, 1997, 2002; and NASS for 1997, 2001, and 2006. In addition, NASS use estimates for propanil only were available for 2006. The NPUD estimates used in the 2001 analysis represent use for 2002, and the NPUD estimates were not included in the 2006–07 analysis. Lastly, the 2006–07 analysis did not include the USEPA use estimates for alachlor and EPTC.

Comparisons of EPest-low and EPest-high total use estimates with the USEPA, NASS, and NPUD data for 1997, 2001–02, and 2006–07 for the seven herbicides are shown in *figures 7A, 7B,* and *7C.* With the exceptions of the EPest-low 2001 estimate for alachlor, the 2007 EPest-low and EPest-high estimates for propanil, and the 2007 EPest-high estimates for trifluralin, EPest and USEPA estimates differed from one another by less than 20 percent. NASS use estimates are not complete national estimates, so they were less than both EPest-low and EPest-high totals, and most 2006 NASS use estimates were a fraction of both USEPA and EPest totals because the number of the crops and states that were surveyed and reported by NASS was reduced in 2006. Overall, the comparisons illustrated in *figure 7* indicate a high level of agreement between EPest totals and both the USEPA and NPUD estimates, although none of these three sources of national estimates is known to be a better estimate of true use than the others.

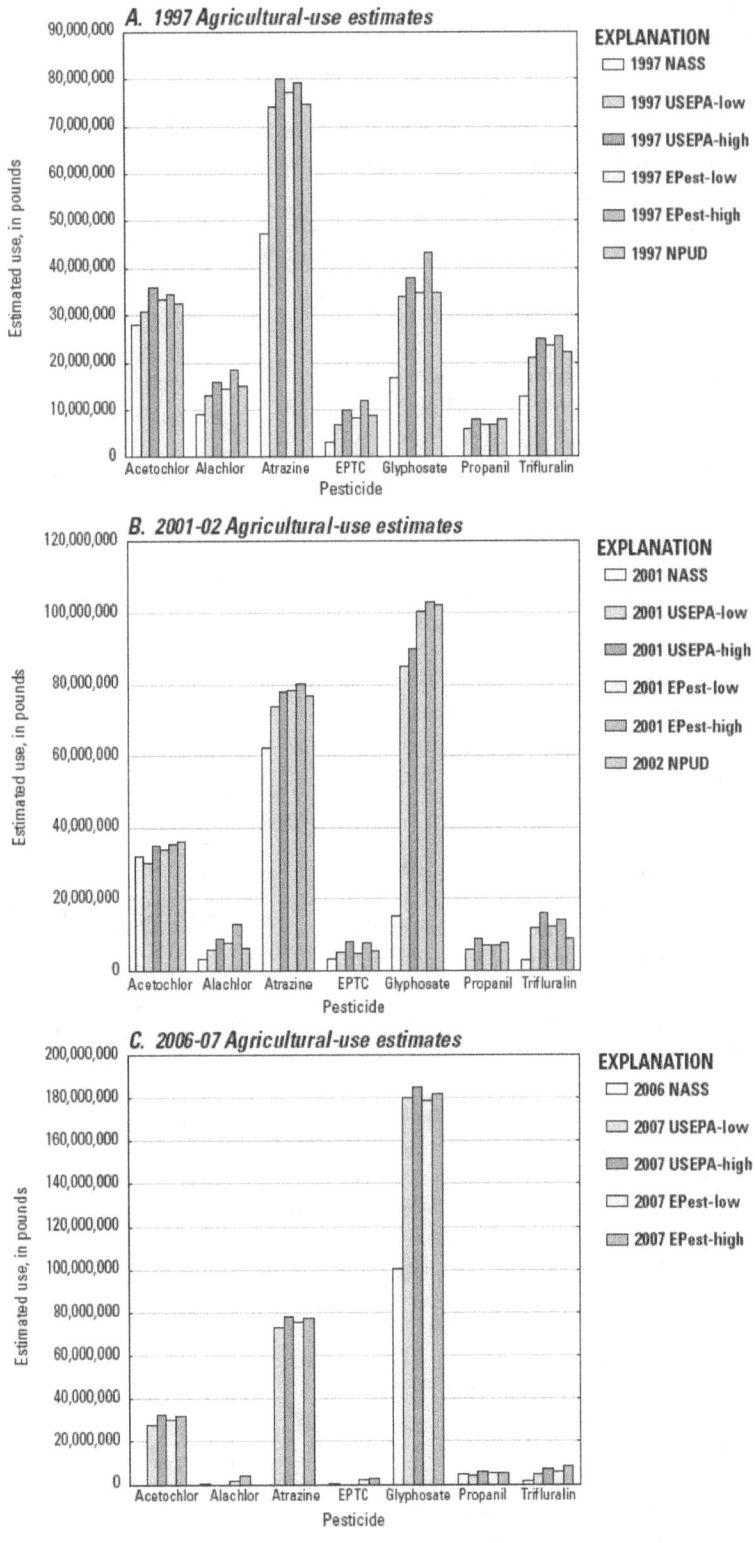

Figure 7. Comparison of EPest-low and EPest-high national total use of selected pesticides with national use estimates from other sources for (*A*) 1997 Agricultural-use estimates, (*B*) 2001–02 Agricultural-use estimates, and (*C*) 2006–07 Agricultural-use estimates. NASS, National Agricultural Statistics Service; USEPA, U.S. Environmental Protection Agency; EPest, estimated pesticide use; National Pesticide Use Database.

Comparisons of EPest and NASS State Estimates

The national comparisons provide an aggregated assessment of how comparable EPest totals are to other published sources. In order to determine how well EPest use estimates represented regional and state level amounts and patterns of pesticide use, a second set of evaluations were made that compared EPest and NASS estimates for (1) state totals for individual pesticides and (2) state totals for individual pesticide-by-crop combinations. The comparisons between EPest and NASS state and state-by-crop estimates were the most controlled evaluations possible.

Comparison of State Total-Use Estimates

State-level comparisons were made for individual pesticides that have four or more estimates for combinations of states, crops, and years common to both EPest and NASS use estimates. Estimates for 33 pesticides and 34 states were compared for one or more years from 1992 through 2006. The pesticides included 24 herbicides, 8 insecticides, and 1 fungicide. Depending on the state and year, estimated state totals represented the sum of a pesticide used on one or more crops, including barley, corn, cotton, peanuts, rice, sorghum, soybeans, spring wheat, sunflowers, tobacco, and winter wheat. For each comparison, the difference between EPest and NASS use estimates was evaluated as the relative error (RE) for EPest relative to NASS estimates, or (EPest − NASS) / NASS, and RE was used to show the distribution of differences in state estimates for each pesticide (*fig. 8*). In *figures 8A* (EPest-low) and *8B* (EPest-high), positive RE values represent EPest totals that were greater than NASS use estimates and negative RE values represent EPest

totals that were less than NASS use estimates. Although differences between EPest and NASS estimates are expressed as proportional errors relative to NASS estimates in order to facilitate clear comparisons to publicly available NASS estimates, neither estimate can be considered a more certain estimate of true values than the other. The number of state-by-year combinations for each pesticide is indicated at the bottom of the plot (*fig.8*). For the different pesticides, the number of state-by-year combinations used in the comparisons ranged from as few as 5 to as many as 443.

Of the 33 pesticides evaluated, less than one-third—10 EPest-low and 8 EPest-high— had median RE values significantly different from zero based on the 95-percent confidence interval on the median RE. For EPest-low, 6 of the 10 pesticides that were significantly different from NASS use estimates tended to have lower estimates compared to NASS (acifluorfen, bentazon, butylate, methomyl, methyl parathion, and propachlor), and the rest (atrazine, fluometuron, nicosulfuron, and propargite) tended to be greater than NASS. Compared to NASS use estimates, seven of the eight significantly different EPest-high totals tended toward overestimation (atrazine, fluometuron, fonofos, metribuzin, nicosulfuron, propargite, and trifluralin), and only one pesticide (methyl parathion) tended toward underestimation. The inter-quartile ranges for both sets of estimates generally were symmetrical for most pesticides, and there was a relatively small proportion of outlying individual values—generally fewer than 10 percent. Several pesticides showed wide confidence intervals around the median, and some had only a small number of estimates to compare, including propachlor and thiobencarb, among others.

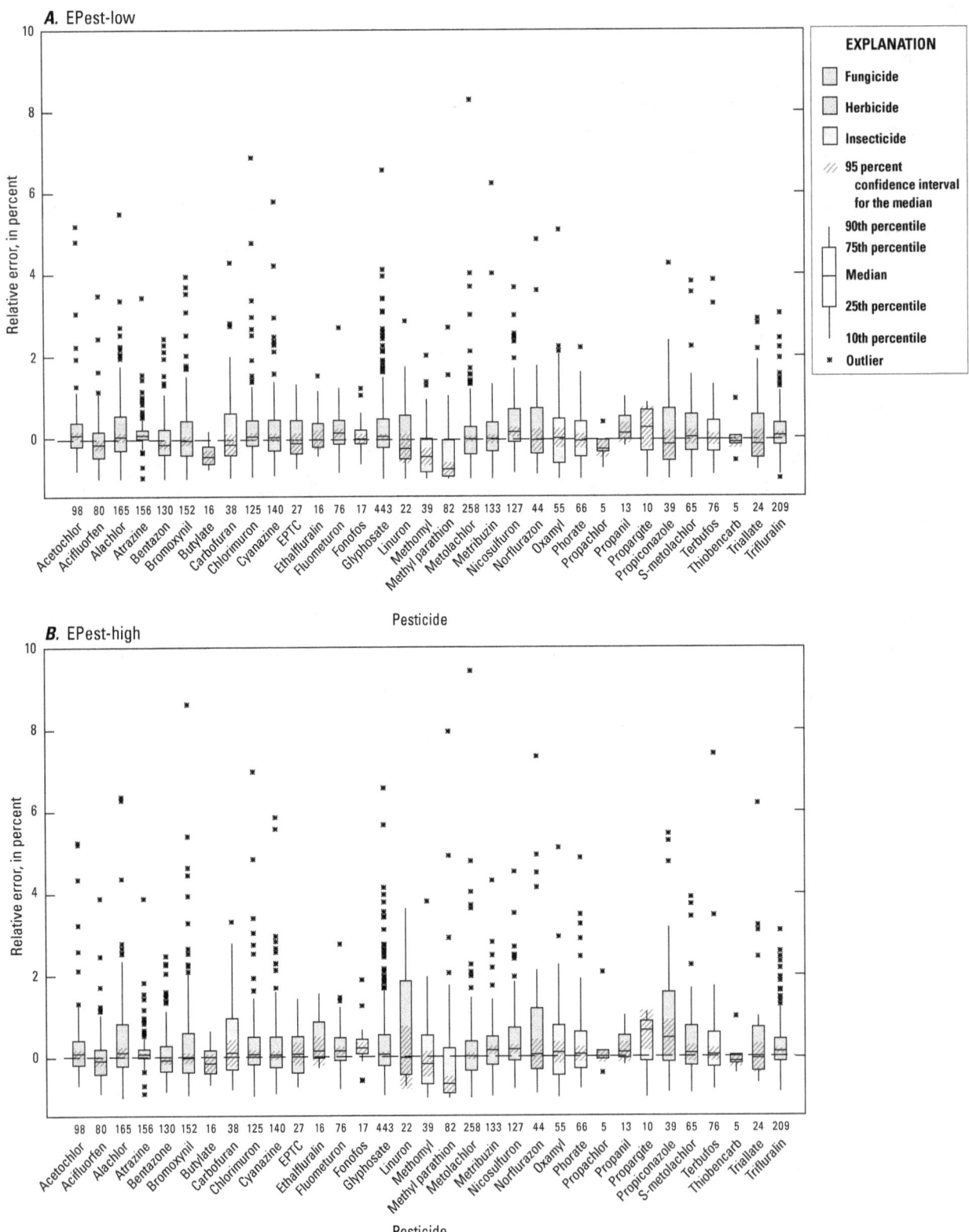

Figure 8.   Distributions of relative error between EPest and National Agricultural Statistics Service (NASS) use estimates. Relative error expressed as (EPEST - NASS)/NASS. Estimated state totals represent the sum of use on one or more crops, including barley, corn, cotton, peanuts, rice, sorghum, soybeans, spring wheat, sunflowers, tobacco, and winter wheat. Numbers for each pesticide represent the number of state-by-year combinations compared.

Comparison of State Estimates for Individual Pesticide-by-Crop Combinations

EPest and NASS use estimates for individual states and crops were compared for selected years from 1992 to 2006, which are the most direct comparisons possible with the data available. The comparisons were limited to pesticide-by-crop combinations that had both EPest and NASS use estimates for at least 10 state-year combinations. This requirement allowed one or more crop comparisons for 29 pesticides, including 21 herbicides, 7 insecticides, and 1 fungicide, for one or more of the following crops: corn, cotton, rice, soybeans, spring wheat, and winter wheat. There were 17 pesticides compared for corn, 13 pesticides for cotton, 9 pesticides for soybeans, 4 pesticides for winter wheat, 4 pesticides for spring wheat, and a single pesticide for rice. Although NASS also reported pesticide-use estimates for other crops included in the all-crops state totals, such as sorghum, tobacco, peanuts, and barley, there were too few estimates for each of these crops to include them in the crop-specific comparisons.

The distribution of RE values for all available state-year combinations for each of the 47 pesticide-by-crop combinations are shown by crop (rice excluded) in *figures 9A–9E* for EPest-low totals and in *figures 10A–10E* for EPest-high totals. The figures show that the range of RE values for EPest-low totals for most pesticide-by-crop combinations was less than for EPest-high totals and contained fewer outliers, indicating that EPest-low totals tended to approximate NASS estimates more accurately than EPest-high totals.

Similarly, more than two-thirds (33 of 48) of EPest-low pesticide-by-crop combinations had median REs that were 15 percent or less, whereas just over half (26 of 48) of the EPest-high totals had median REs that were less 15 percent or less (*tables 3* and *4*). Of the 15 EPest-low pesticide-by-crop combinations that had median REs that differed by 15 percent or more, 13 pesticide crop-combinations were less than NASS use estimates and 2 pesticide-by-crop combinations were greater than NASS use estimates (*table 3*). There were 21 EPest-high pesticide-by-crop combinations that had median REs greater than 15 percent, with 13 combinations greater than NASS use estimates and 8 that were less (*table 4*). These results were consistent with the aggregated state total comparisons presented previously, and overall, these comparisons indicated a reasonable agreement between EPest and NASS use estimates, with somewhat better agreement for EPest-low than high estimates. Nevertheless, some pesticide-by-crop combinations showed substantial differences in the estimates for specific states and years.

A combination of statistical tests were used to compare EPest and NASS use estimates for the pesticide-by-crop combinations. The Wilcoxon signed rank sum test (Conover, 1980; Lehmann, 1975) was used to further evaluate differences between magnitudes of EPest and NASS annual use estimates for each pesticide-by-crop combination with sufficient state-year combinations. This non-parametric test evaluates whether the median difference between paired estimates is significantly different than zero, where significance was assigned to a probability (p) of less than 0.05 (two-tailed test). Comparisons that are not statistically significant can indicate agreement between estimates or also can indicate variability in the sample too great to establish significant differences. To help assess the degree of correlation between two ranked pairs of estimates, the Spearman rank correlation coefficient (r) was used, where values range from 0 to 1, and 1 indicates perfect agreement between estimates. The p-value from the Wilcoxon test, the Spearman correlation coefficient (r), the median RE, and the number of state/year combinations used in the evaluations of the comparisons to NASS use estimates are shown for each pesticide and crop combination in *table 3* for EPest-low and *table 4* for EPest-high.

The strongest agreement between estimates is indicated by statistically insignificant p-values, correlation coefficients approaching 1, and a low median and range for RE values. Pesticides evaluated in this study that met these criteria included acetochlor, cyanzine, and terbufos use estimates for corn, as well as chlorimuron and bentazon use estimates for soybeans. Some estimate comparisons had significantly different medians, but still showed strong correlation and a low RE value; examples include estimates for atrazine and metolachlor use for corn and trifluralin use estimates for cotton. Poor agreement between estimates was indicated by large RE values and low correlation coefficients for both significant and insignificant comparisons of medians. A small sample size can reduce the power of the tests, however, and smaller sample sizes were often associated with the lower correlation coefficients among these comparisons, particularly when RE values were greater than 0.15.

More than half of the comparisons of pesticide-by-crop combinations had RE values less than 0.15, and the majority of these comparisons were not significantly different and had correlation coefficients greater than 0.75. Of the 48 pesticide-by-crop combinations with 10 or more state-by-year combinations, 12 of the EPest-low pesticide-by-crop totals and 17 of the EPest-high totals significantly differed (p < 0.05) from the NASS use estimates. Of the comparisons with significant differences, two-thirds or more of the pesticide-by-crop combinations had correlation coefficients greater than 0.75, especially when comparisons had RE values of 0.15 or less. Comparisons that did not have significant differences tended to have lower RE values than comparisons that had significant differences. Nevertheless, about a quarter of all the comparisons had RE values greater than 0.15, but did not have significant differences. All of these had sample numbers less than 40, and most had fewer than 20 samples for comparison. Also, most had correlation coefficients less than 0.75, which demonstrates the importance of having a sample number large enough to achieve a good comparison of estimates.

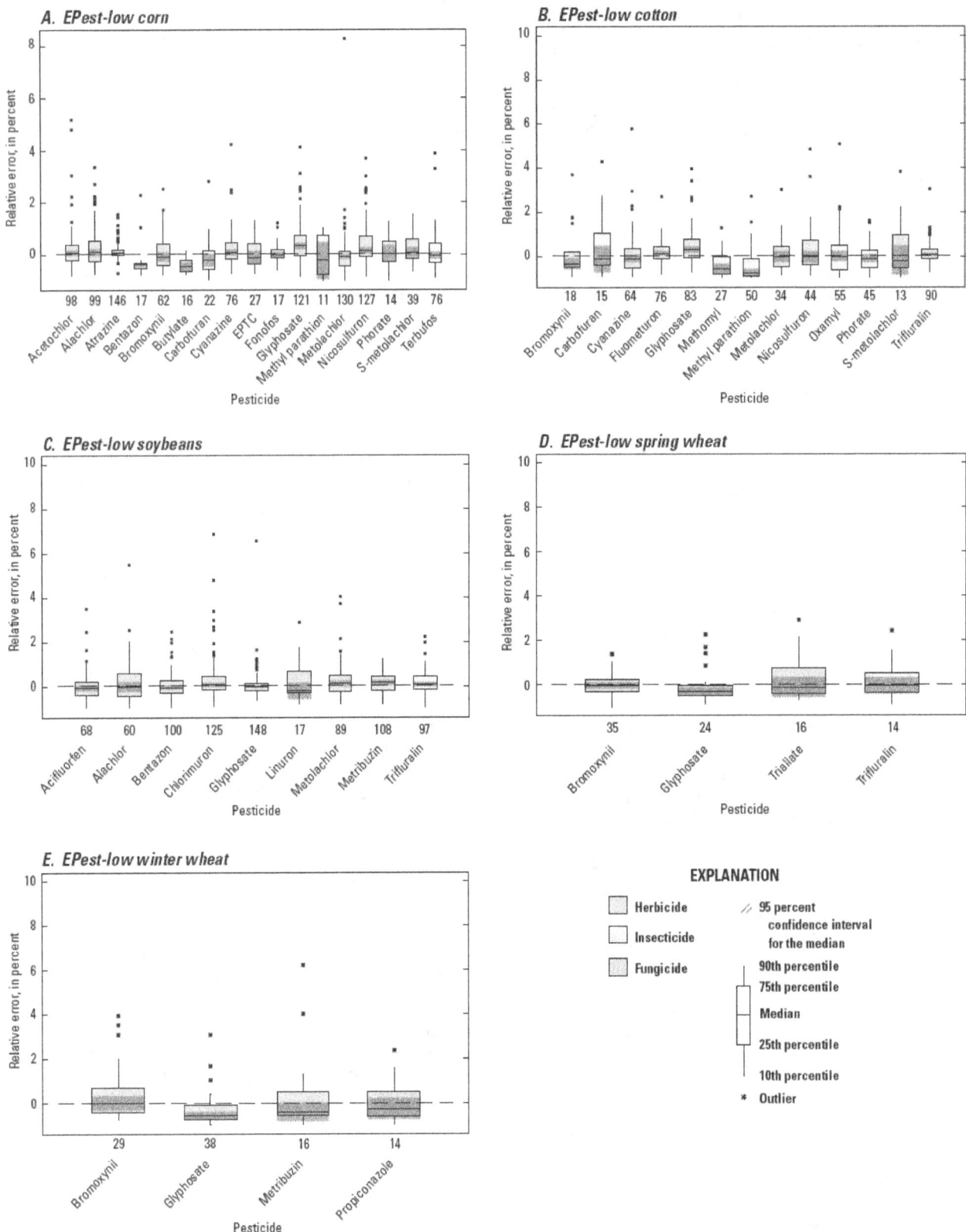

Figure 9.   Distribution of relative error between EPest-low and National Agricultural Statistics Service (NASS) use estimates for (*A*) corn, (*B*) cotton, (*C*) soybeans, (*D*) spring wheat, and (*E*) winter wheat. Relative error expressed as (EPEST - NASS)/NASS.

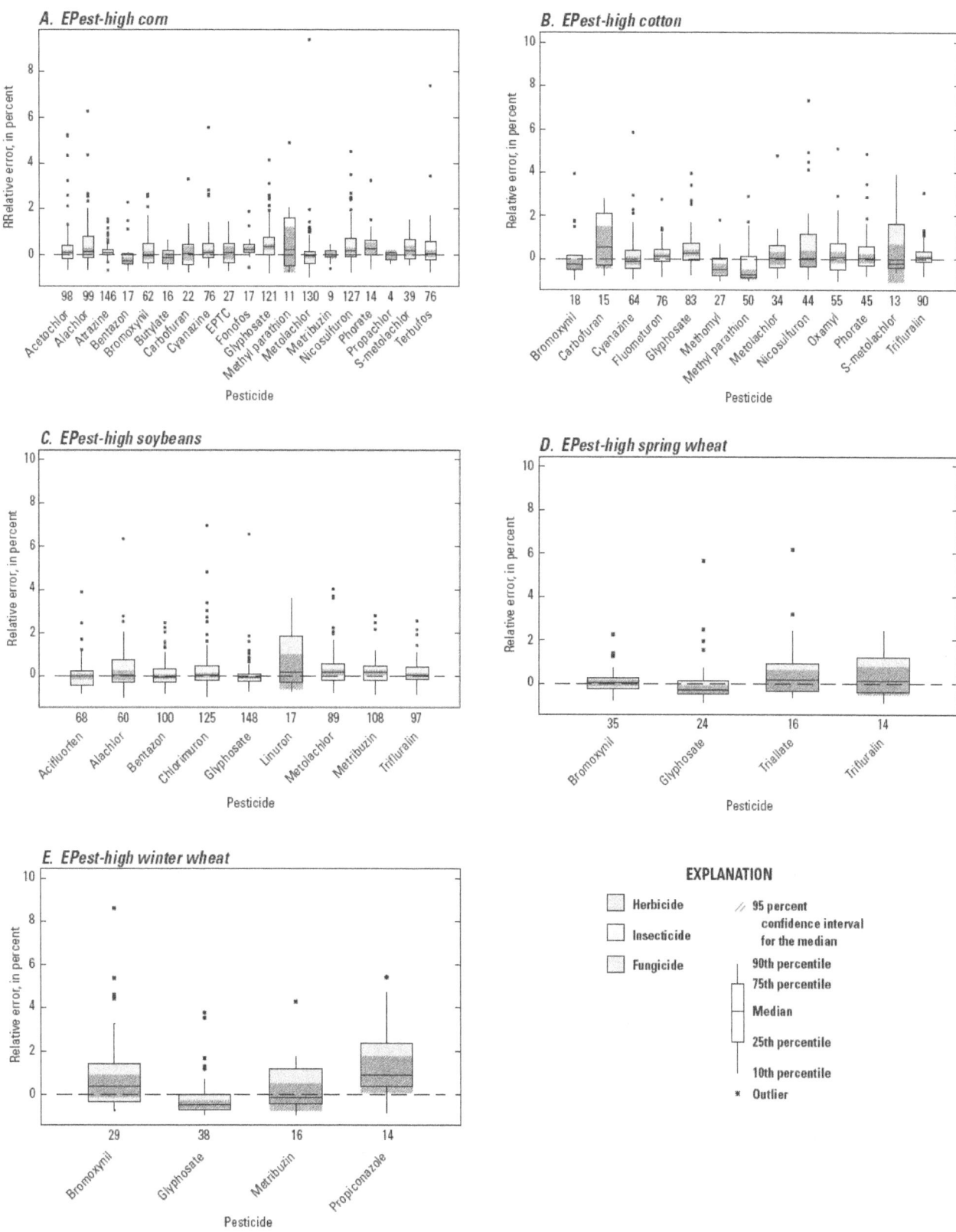

Figure 10.   Distribution of relative error between EPest-high and National Agricultural Statistics Service (NASS) use estimates for (*A*) corn, (*B*) cotton, (*C*) soybeans, (*D*) spring wheat, and (*E*) winter wheat. Relative error expressed as (EPEST - NASS)/NASS.

Table 3. Summary of statistics from comparison of EPest-low and National Agricultural Statistics Service (NASS) pesticide-by-crop estimates.

[Abbreviations: N, number of estimates compared; P, probability of significance; >, greater than; <, less than; —, no data]

Pesticide	Type	Corn N	Corn Wilcoxon signed rank, P (two-tailed)	Corn Wilcoxon signed rank, P (NASS >EPest)	Corn Wilcoxon signed rank, P (NASS <EPest)	Corn Median relative error	Corn Spearman correlation coefficient	Cotton N	Cotton Wilcoxon signed rank, P (two-tailed)	Cotton Wilcoxon signed rank, P (NASS >EPest)	Cotton Wilcoxon signed rank, P (NASS <EPest)	Cotton Median relative error	Cotton Spearman correlation coefficient
Acetochlor	Herbicide	98	0.19	0.90	0.10	0.07	0.93	—	—	—	—	—	—
Aciflurofen	Herbicide	—	—	—	—	—	—	—	—	—	—	—	—
Alachlor	Herbicide	99	0.13	0.94	0.06	0.08	0.83	—	—	—	—	—	—
Atrazine	Herbicide	146	0.00	1.00	0.00	0.07	0.97	—	—	—	—	—	—
Bentazon	Herbicide	17	0.01	0.01	0.99	(0.39)	0.42	—	—	—	—	—	—
Bromoxynil	Herbicide	62	0.29	0.15	0.86	(0.09)	0.85	18	0.12	0.06	0.95	(0.36)	0.71
Butylate	Herbicide	16	0.00	0.00	1.00	(0.47)	0.91	—	—	—	—	—	—
Carbofuran	Insecticide	22	0.18	0.09	0.92	(0.22)	0.68	15	0.56	0.74	0.28	(0.07)	0.73
Chlorimuron	Herbicide	—	—	—	—	—	—	—	—	—	—	—	—
Cyanzine	Herbicide	76	0.20	0.90	0.10	0.07	0.92	64	0.65	0.32	0.68	(0.11)	0.82
EPTC	Herbicide	27	0.24	0.12	0.88	(0.13)	0.60	—	—	—	—	—	—
Fluometuron	Herbicide	—	—	—	—	—	—	76	0.00	1.00	0.00	0.12	0.93
Fonofos	Insecticide	17	0.85	0.59	0.43	(0.03)	0.72	—	—	—	—	—	—
Glyphosate	Herbicide	121	0.00	1.00	0.00	0.34	0.78	83	0.00	1.00	0.00	0.30	0.93
Linuron	Herbicide	—	—	—	—	—	—	—	—	—	—	—	—
Methomyl	Insecticide	—	—	—	—	—	—	27	0.01	0.00	1.00	(0.61)	0.76
Methyl parathion	Insecticide	11	0.15	0.07	0.94	(0.24)	0.32	50	0.00	0.00	1.00	(0.78)	0.47
Metolachlor	Herbicide	130	0.00	0.00	1.00	(0.10)	0.87	34	0.30	0.15	0.85	(0.04)	0.58
Metribuzin	Herbicide	—	—	—	—	—	—	—	—	—	—	—	—
Nicosulfuron	Herbicide	127	0.00	1.00	0.00	0.14	0.84	—	—	—	—	—	—
Norflurazon	Herbicide	—	—	—	—	—	—	44	0.48	0.24	0.76	(0.06)	0.78
Oxamyl	Insecticide	—	—	—	—	—	—	55	0.47	0.23	0.77	0.02	0.68
Phorate	Insecticide	14	0.86	0.60	0.43	0.03	0.50	45	0.62	0.31	0.69	(0.12)	0.80
Propanil	Herbicide	—	—	—	—	—	—	—	—	—	—	—	—
Propiconazole	Fungicide	—	—	—	—	—	—	—	—	—	—	—	—
S-Metolachlor	Herbicide	39	0.06	0.97	0.03	0.08	0.90	13	0.41	0.21	0.81	(0.21)	0.37
Terbufos	Insecticide	76	0.29	0.15	0.86	(0.05)	0.83	—	—	—	—	—	—
Triallate	Herbicide	—	—	—	—	—	—	—	—	—	—	—	—
Trifluralin	Herbicide	—	—	—	—	—	—	90	0.08	0.96	0.04	0.06	0.96

Table 3. Summary of statistics from comparison of EPest-low and National Agricultural Statistics Service (NASS) pesticide-by-crop estimates.—Continued

[Abbreviations: N, number of estimates compared; P, probability of significance; >, greater than; <, less than; —, no data]

Pesticide	Type	Soybeans						Spring wheat					
		N	Wilcoxon signed rank, P (two-tailed)	Wilcoxon signed rank, P (NASS >EPest)	Wilcoxon signed rank, P (NASS <EPest)	Median relative error	Spearman correlation coefficient	N	Wilcoxon signed rank, P (two-tailed)	Wilcoxon signed rank, P (NASS >EPest)	Wilcoxon signed rank, P (NASS <EPest)	Median relative error	Spearman correlation coefficient
Acetochlor	Herbicide	—	—	—	—	—	—	—	—	—	—	—	—
Aciflurofen	Herbicide	68	0.06	0.03	0.97	(0.10)	0.81	—	—	—	—	—	—
Alachlor	Herbicide	60	0.85	0.42	0.58	(0.04)	0.69	—	—	—	—	—	—
Atrazine	Herbicide	—	—	—	—	—	—	—	—	—	—	—	—
Bentazon	Herbicide	100	0.22	0.11	0.89	(0.08)	0.89	—	—	—	—	—	—
Bromoxynil	Herbicide	—	—	—	—	—	—	35	0.68	0.34	0.67	(0.04)	0.85
Butylate	Herbicide	—	—	—	—	—	—	—	—	—	—	—	—
Carbofuran	Insecticide	—	—	—	—	—	—	—	—	—	—	—	—
Chlorimuron	Herbicide	125	0.16	0.92	0.08	0.03	0.82	—	—	—	—	—	—
Cyanzine	Herbicide	—	—	—	—	—	—	—	—	—	—	—	—
EPTC	Herbicide	—	—	—	—	—	—	—	—	—	—	—	—
Fluometuron	Herbicide	—	—	—	—	—	—	—	—	—	—	—	—
Fonofos	Insecticide	—	—	—	—	—	—	—	—	—	—	—	—
Glyphosate	Herbicide	148	0.06	0.03	0.97	(0.06)	0.97	24	0.00	0.00	1.00	(0.31)	0.85
Linuron	Herbicide	17	0.68	0.34	0.68	(0.25)	0.85	—	—	—	—	—	—
Methomyl	Insecticide	—	—	—	—	—	—	—	—	—	—	—	—
Methyl parathion	Insecticide	—	—	—	—	—	—	—	—	—	—	—	—
Metolachlor	Herbicide	89	0.08	0.96	0.04	0.08	0.76	—	—	—	—	—	—
Metribuzin	Herbicide	108	0.17	0.92	0.08	0.12	0.81	—	—	—	—	—	—
Nicosulfuron	Herbicide	—	—	—	—	—	—	—	—	—	—	—	—
Norflurazon	Herbicide	—	—	—	—	—	—	—	—	—	—	—	—
Oxamyl	Insecticide	—	—	—	—	—	—	—	—	—	—	—	—
Phorate	Insecticide	—	—	—	—	—	—	—	—	—	—	—	—
Propanil	Herbicide	—	—	—	—	—	—	—	—	—	—	—	—
Propiconazole	Fungicide	—	—	—	—	—	—	—	—	—	—	—	—
S-Metolachlor	Herbicide	—	—	—	—	—	—	—	—	—	—	—	—
Terbufos	Insecticide	—	—	—	—	—	—	—	—	—	—	—	—
Triallate	Herbicide	—	—	—	—	—	—	16	0.78	0.39	0.63	(0.13)	0.53
Trifluralin	Herbicide	97	0.12	0.94	0.06	0.03	0.91	14	0.86	0.43	0.60	(0.06)	0.74

Table 3. Summary of statistics from comparison of EPest-low and National Agricultural Statistics Service (NASS) pesticide-by-crop estimates.—Continued

[Abbreviations: N, number of estimates compared; P, probability of significance; >, greater than; <, less than; –, no data]

Pesticide	Type	Winter wheat						Rice					
		N	Wilcoxon signed rank, P (two-tailed)	Wilcoxon signed rank, P (NASS >EPest)	Wilcoxon signed rank, P (NASS <EPest)	Median relative error	Spearman correlation coefficient	N	Wilcoxon signed rank, P (two-tailed)	Wilcoxon signed rank, P (NASS >EPest)	Wilcoxon signed rank, P (NASS <EPest)	Median relative error	Spearman correlation coefficient
Acetochlor	Herbicide	–	–	–	–	–	–	–	–	–	–	–	–
Aciflurofen	Herbicide	–	–	–	–	–	–	–	–	–	–	–	–
Alachlor	Herbicide	–	–	–	–	–	–	–	–	–	–	–	–
Atrazine	Herbicide	–	–	–	–	–	–	–	–	–	–	–	–
Bentazon	Herbicide	–	–	–	–	–	–	–	–	–	–	–	–
Bromoxynil	Herbicide	29	0.70	0.66	0.34	0.00	0.60	–	–	–	–	–	–
Butylate	Herbicide	–	–	–	–	–	–	–	–	–	–	–	–
Carbofuran	Insecticide	–	–	–	–	–	–	–	–	–	–	–	–
Chlorimuron	Herbicide	–	–	–	–	–	–	–	–	–	–	–	–
Cyanzine	Herbicide	–	–	–	–	–	–	–	–	–	–	–	–
EPTC	Herbicide	–	–	–	–	–	–	–	–	–	–	–	–
Fluometuron	Herbicide	–	–	–	–	–	–	–	–	–	–	–	–
Fonofos	Insecticide	–	–	–	–	–	–	–	–	–	–	–	–
Glyphosate	Herbicide	38	0.00	0.00	1.00	(0.58)	0.58	–	–	–	–	–	–
Linuron	Herbicide	–	–	–	–	–	–	–	–	–	–	–	–
Methomyl	Insecticide	–	–	–	–	–	–	–	–	–	–	–	–
Methyl parathion	Insecticide	–	–	–	–	–	–	–	–	–	–	–	–
Metolachlor	Herbicide	–	–	–	–	–	–	–	–	–	–	–	–
Metribuzin	Herbicide	16	0.53	0.26	0.75	(0.41)	0.11	–	–	–	–	–	–
Nicosulfuron	Herbicide	–	–	–	–	–	–	–	–	–	–	–	–
Norflurazon	Herbicide	–	–	–	–	–	–	–	–	–	–	–	–
Oxamyl	Insecticide	–	–	–	–	–	–	–	–	–	–	–	–
Phorate	Insecticide	–	–	–	–	–	–	–	–	–	–	–	–
Propanil	Herbicide	–	–	–	–	–	–	13	0.68	0.68	0.34	0.11	0.94
Propiconazole	Fungicide	14	0.81	0.40	0.62	(0.27)	0.78	–	–	–	–	–	–
S-Metolachlor	Herbicide	–	–	–	–	–	–	–	–	–	–	–	–
Terbufos	Insecticide	–	–	–	–	–	–	–	–	–	–	–	–
Triallate	Herbicide	–	–	–	–	–	–	–	–	–	–	–	–
Trifluralin	Herbicide	–	–	–	–	–	–	–	–	–	–	–	–

Table 4. Summary of statistics from comparison of EPest-high and National Agricultural Statistics Service (NASS) pesticide-by-crop estimates.

[Abbreviations: N, number of estimates compared; P, probability of significance; <, less than; >, greater than; –, no data]

Pesticide	Type	Corn						Cotton					
		N	Wilcoxon signed rank, P (two-tailed)	Wilcoxon signed rank, P (NASS > EPest)	Wilcoxon signed rank, P (NASS < EPest)	Median relative error	Spearman correlation coefficient	N	Wilcoxon signed rank, P (two-tailed)	Wilcoxon signed rank, P (NASS > EPest)	Wilcoxon signed rank, P (NASS < EPest)	Median relative error	Spearman correlation coefficient
Acetochlor	Herbicide	98	0.07	0.96	0.36	0.08	0.93	–	–	–	–	–	–
Aciflurofen	Herbicide	–	–	–	–	–	–	–	–	–	–	–	–
Alachlor	Herbicide	99	0.01	1.00	0.00	0.13	0.82	–	–	–	–	–	–
Atrazine	Herbicide	146	0.00	1.00	0.00	0.07	0.97	–	–	–	–	–	–
Bentazon	Herbicide	17	0.04	0.22	0.98	(0.30)	0.34	–	–	–	–	–	–
Bromoxynil	Herbicide	62	0.96	0.53	0.48	(0.03)	0.85	18	0.15	0.08	0.93	(0.26)	0.74
Butylate	Herbicide	16	0.19	0.10	0.91	(0.16)	0.81	–	–	–	–	–	–
Carbofuran	Insecticide	22	0.82	0.41	0.60	0.05	0.74	15	0.30	0.86	0.15	0.55	0.63
Chlorimuron	Herbicide	–	–	–	–	–	–	–	–	–	–	–	–
Cyanzine	Herbicide	76	0.09	0.95	0.05	0.08	0.92	64	0.92	0.54	0.46	(0.08)	0.83
EPTC	Herbicide	27	0.71	0.35	0.65	0.07	0.56	–	–	–	–	–	–
Fluometuron	Herbicide	–	–	–	–	–	–	76	0.00	1.00	0.00	0.14	0.93
Fonofos	Insecticide	17	0.05	0.98	0.03	0.21	0.73	–	–	–	–	–	–
Glyphosate	Herbicide	121	0.00	1.00	0.00	0.34	0.79	83	0.00	1.00	0.00	0.30	0.92
Linuron	Herbicide	–	–	–	–	–	–	–	–	–	–	–	–
Methomyl	Insecticide	–	–	–	–	–	–	27	0.05	0.27	0.97	(0.46)	0.74
Methyl parathion	Insecticide	11	1.00	0.52	0.52	0.20	0.16	50	0.00	0.00	1.00	(0.69)	0.52
Metolachlor	Herbicide	130	0.01	0.00	1.00	(0.07)	0.88	34	0.99	0.49	0.51	0.05	0.65
Metribuzin	Herbicide	–	–	–	–	–	–	–	–	–	–	–	–
Nicosulfuron	Herbicide	127	0.00	1.00	0.00	0.17	0.84	–	–	–	–	–	–
Norflurazon	Herbicide	–	–	–	–	–	–	44	0.38	0.81	0.19	0.05	0.74
Oxamyl	Insecticide	–	–	–	–	–	–	55	0.65	0.68	0.32	0.10	0.64
Phorate	Insecticide	14	0.05	0.98	0.02	0.28	0.59	45	0.22	0.89	0.11	0.05	0.71
Propanil	Herbicide	–	–	–	–	–	–	–	–	–	–	–	–
Propiconazole	Fungicide	–	–	–	–	–	–	–	–	–	–	–	–
S-Metolachlor	Herbicide	39	0.04	0.98	0.02	0.16	0.91	13	1.00	0.53	0.50	(0.17)	0.37
Terbufos	Insecticide	76	0.51	0.75	0.25	0.04	0.83	–	–	–	–	–	–
Triallate	Herbicide	–	–	–	–	–	–	–	–	–	–	–	–
Trifluralin	Herbicide	–	–	–	–	–	–	90	0.03	0.99	0.01	0.10	0.95

Table 4.   Summary of statistics from comparison of EPest-high and National Agricultural Statistics Service (NASS) pesticide-by-crop estimates.—Continued

[Abbreviations: N, number of estimates compared; P, probability of significance; <, less than; >, greater than; —, no data]

Pesticide	Type	Soybeans							Spring wheat					
		N	Wilcoxon signed rank, P (two-tailed)	Wilcoxon signed rank, P (NASS > EPest)	Wilcoxon signed rank, P (NASS < EPest)	Median relative error	Spearman correlation coefficient	N	Wilcoxon signed rank, P (two-tailed)	Wilcoxon signed rank, P (NASS > EPest)	Wilcoxon signed rank, P (NASS < EPest)	Median relative error	Spearman correlation coefficient
Acetochlor	Herbicide	—	—	—	—	—	—	—	—	—	—	—	—
Aciflurofen	Herbicide	68	0.29	0.15	0.85	(0.01)	0.81	—	—	—	—	—	—
Alachlor	Herbicide	60	0.29	0.86	0.14	0.05	0.70	—	—	—	—	—	—
Atrazine	Herbicide	—	—	—	—	—	—	—	—	—	—	—	—
Bentazon	Herbicide	100	0.58	0.29	0.71	(0.01)	0.89	—	—	—	—	—	—
Bromoxynil	Herbicide	—	—	—	—	—	—	35	0.98	0.49	0.51	0.02	0.83
Butylate	Herbicide	—	—	—	—	—	—	—	—	—	—	—	—
Carbofuran	Insecticide	—	—	—	—	—	—	—	—	—	—	—	—
Chlorimuron	Herbicide	125	0.82	0.96	0.04	0.06	0.82	—	—	—	—	—	—
Cyanzine	Herbicide	—	—	—	—	—	—	—	—	—	—	—	—
EPTC	Herbicide	—	—	—	—	—	—	—	—	—	—	—	—
Fluometuron	Herbicide	—	—	—	—	—	—	—	—	—	—	—	—
Fonofos	Insecticide	—	—	—	—	—	—	—	—	—	—	—	—
Glyphosate	Herbicide	148	0.10	0.05	0.95	(0.05)	0.97	24	0.01	0.00	1.00	(0.28)	0.84
Linuron	Herbicide	17	0.64	0.69	0.32	0.22	0.80	—	—	—	—	—	—
Methomyl	Insecticide	—	—	—	—	—	—	—	—	—	—	—	—
Methyl parathion	Insecticide	—	—	—	—	—	—	—	—	—	—	—	—
Metolachlor	Herbicide	89	0.00	1.00	0.00	0.19	0.75	—	—	—	—	—	—
Metribuzin	Herbicide	108	0.02	0.99	0.01	0.18	0.80	—	—	—	—	—	—
Nicosulfuron	Herbicide	—	—	—	—	—	—	—	—	—	—	—	—
Norflurazon	Herbicide	—	—	—	—	—	—	—	—	—	—	—	—
Oxamyl	Insecticide	—	—	—	—	—	—	—	—	—	—	—	—
Phorate	Insecticide	—	—	—	—	—	—	—	—	—	—	—	—
Propanil	Herbicide	—	—	—	—	—	—	—	—	—	—	—	—
Propiconazole	Fungicide	—	—	—	—	—	—	—	—	—	—	—	—
S-Metolachlor	Herbicide	—	—	—	—	—	—	—	—	—	—	—	—
Terbufos	Insecticide	—	—	—	—	—	—	—	—	—	—	—	—
Triallate	Herbicide	—	—	—	—	—	—	16	0.82	0.06	0.41	0.15	0.53
Trifluralin	Herbicide	97	0.04	0.98	0.02	0.07	0.91	14	0.86	0.60	0.43	0.14	0.74

Table 4. Summary of statistics from comparison of EPest-high and National Agricultural Statistics Service (NASS) pesticide-by-crop estimates.—Continued

[**Abbreviations**: N, number of estimates compared; P, probability of significance; <, less than; >, greater than; —, no data]

Pesticide	Type	Winter wheat						Rice					
		N	Wilcoxon signed rank, P (two-tailed)	Wilcoxon signed rank, P (NASS > EPest)	Wilcoxon signed rank, P (NASS < EPest)	Median relative error	Spearman correlation coefficient	N	Wilcoxon signed rank, P (two-tailed)	Wilcoxon signed rank, P (NASS > EPest)	Wilcoxon signed rank, P (NASS < EPest)	Median relative error	Spearman correlation coefficient
Acetochlor	Herbicide	—	—	—	—	—	—	—	—	—	—	—	—
Aciflurofen	Herbicide	—	—	—	—	—	—	—	—	—	—	—	—
Alachlor	Herbicide	—	—	—	—	—	—	—	—	—	—	—	—
Atrazine	Herbicide	—	—	—	—	—	—	—	—	—	—	—	—
Bentazon	Herbicide	—	—	—	—	—	—	—	—	—	—	—	—
Bromoxynil	Herbicide	29	0.23	0.89	0.11	0.39	0.40	—	—	—	—	—	—
Butylate	Herbicide	—	—	—	—	—	—	—	—	—	—	—	—
Carbofuran	Insecticide	—	—	—	—	—	—	—	—	—	—	—	—
Chlorimuron	Herbicide	—	—	—	—	—	—	—	—	—	—	—	—
Cyanzine	Herbicide	—	—	—	—	—	—	—	—	—	—	—	—
EPTC	Herbicide	—	—	—	—	—	—	—	—	—	—	—	—
Fluometuron	Herbicide	—	—	—	—	—	—	—	—	—	—	—	—
Fonofos	Insecticide	—	—	—	—	—	—	—	—	—	—	—	—
Glyphosate	Herbicide	38	0.00	0.00	1.00	(0.46)	0.58	—	—	—	—	—	—
Linuron	Herbicide	—	—	—	—	—	—	—	—	—	—	—	—
Methomyl	Insecticide	—	—	—	—	—	—	—	—	—	—	—	—
Methyl parathion	Insecticide	—	—	—	—	—	—	—	—	—	—	—	—
Metolachlor	Herbicide	—	—	—	—	—	—	—	—	—	—	—	—
Metribuzin	Herbicide	16	0.60	0.72	0.30	(0.13)	0.16	—	—	—	—	—	—
Nicosulfuron	Herbicide	—	—	—	—	—	—	—	—	—	—	—	—
Norflurazon	Herbicide	—	—	—	—	—	—	—	—	—	—	—	—
Oxamyl	Insecticide	—	—	—	—	—	—	—	—	—	—	—	—
Phorate	Insecticide	—	—	—	—	—	—	—	—	—	—	—	—
Propanil	Herbicide	—	—	—	—	—	—	13	0.68	0.68	0.34	0.11	0.95
Propiconazole	Fungicide	14	0.00	1.00	0.00	0.92	0.65	—	—	—	—	—	—
S-Metolachlor	Herbicide	—	—	—	—	—	—	—	—	—	—	—	—
Terbufos	Insecticide	—	—	—	—	—	—	—	—	—	—	—	—
Triallate	Herbicide	—	—	—	—	—	—	—	—	—	—	—	—
Trifluralin	Herbicide	—	—	—	—	—	—	—	—	—	—	—	—

Comparisons of EPest-low tended to show stronger correlation to NASS use estimates than EPest-high and also had a greater number of RE values less than 0.15, which, along with fewer significant differences between medians, indicated that EPest-low totals better approximated NASS use estimates than EPest-high overall. In general, however, the majority of the comparisons of estimates showed agreement, although low sample size limited the power of the tests for some pesticide-by-crop combinations.

Comparisons of EPest-low and EPest-high crop-pesticide combinations with NASS use estimates were further examined to evaluate differences between the estimates. These comparisons provide an understanding of the types and degrees of differences between EPest and NASS estimates and how the statistical tests summarize them.

Herbicide Estimate Comparisons between EPest and NASS

Statistically significant differences in median estimates between the methods are important to understand because they can provide information about similarities and differences in the estimates. One or both EPest medians for 11 of the 21 herbicides were significantly different than NASS median use estimates (*tables 3* and *4*). For six of these herbicides—atrazine, bentazon, fluometuron, glyphosate, metolachlor, and nicosulfuron—both EPest-low and EPest-high medians differed significantly from NASS median use estimates. In addition, EPest-high (but not EPest-low) medians for alachlor,

metribuzin, *S*-metolachlor, and trifluralin were significantly different from NASS median use estimates, and EPest-low (but not EPest-high) medians for butylate were significantly different from NASS median use estimates. Use estimates for more than one crop were compared for some pesticides, such as metolachlor and bentazon, and both EPest medians (low and high) were significantly different from NASS median use estimates for some but not all of the crops that were compared. For example, EPest-low and EPest-high bentazon medians were significantly different than NASS median use estimates for corn but not soybeans.

Examining the data and statistical results of the pesticide-by-crop comparisons can help to better assess and understand how well the EPest method approximated current NASS pesticide-use estimates. The following sections present the data graphically and discuss the results of the statistical tests for a selection of the pesticide-by-crop combinations that showed significant differences for one or both methods. For all pesticide-by-crop combinations presented, two plots are shown: (1) a scatterplot of EPest-low and NASS state pesticide-use totals for the years compared (only plots of EPest-low estimates were used because they are similar to the EPest-high versions of the scatterplots) and (2) a plot of differences between EPest estimates and NASS state pesticide-use estimates on a common scale, organized by USDA Farm Production Regions. Because their boundaries conform to state boundaries, Farm Production Regions (*fig. 11*) were selected rather than the USDA Farm Resource Regions that were used to calculate EPest regional rates.

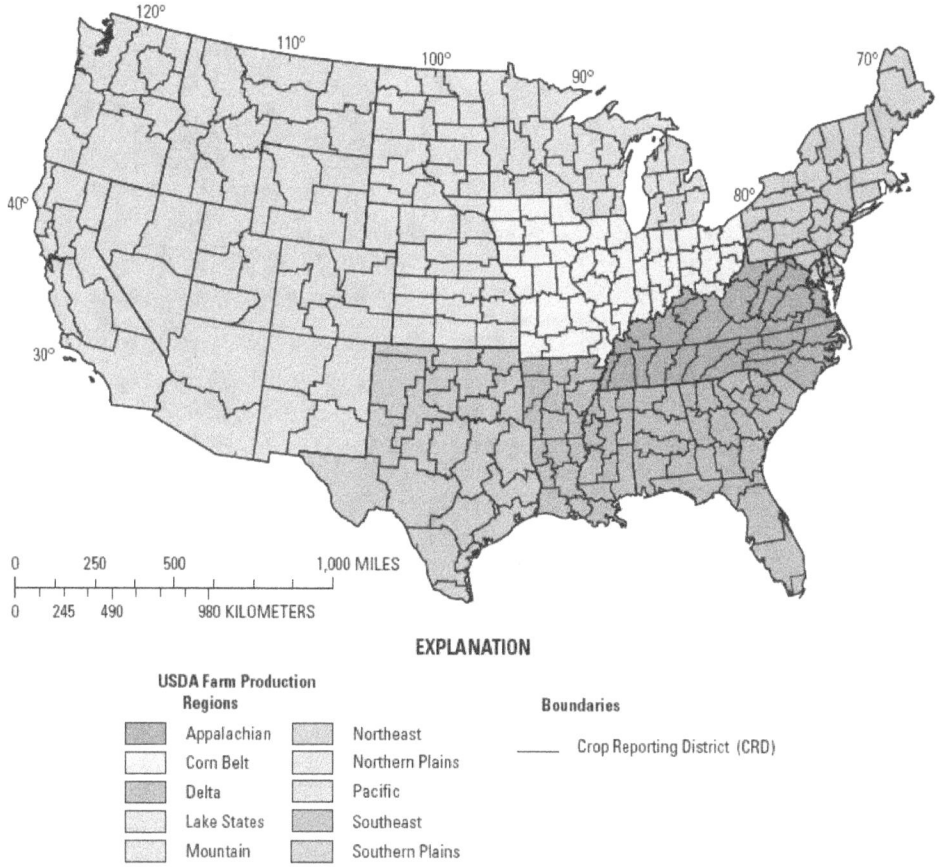

Figure 11.   U.S. Department of Agriculture Farm Production Regions.

Alachlor

For 19 states and most of the years from 1992 through 2003, 99 EPest-low and EPest-high estimates of alachlor use on corn were compared with NASS estimates. Only EPest-high estimates significantly differed (p <0.05) from NASS use estimates, but both EPest totals tended to be greater than NASS totals. The medians of the RE distributions comparing EPest-low and EPest-high to NASS estimates were 8 and 13 percent greater, respectively, indicating a general tendency for EPest estimates to be greater than NASS estimates. Correlation coefficients for EPest-low and NASS comparisons

were 0.83 and were 0.82 for EPest-high. The relation between EPest-low and NASS estimates for alachlor is shown in *figure 12A*, and the differences between NASS estimates and both EPest-low and EPest-high are shown by region and state in *figure 12B*.

The majority of EPest-low and EPest-high estimates differed from NASS use estimates by less than a factor of two (*fig. 12B*), and most EPest and NASS use estimates followed similar trends use for the years compared. Of the approximately 20 percent (20 of 99) of EPest-high estimates that were more than double the NASS estimate, most were in the Corn Belt and Lake States regions.

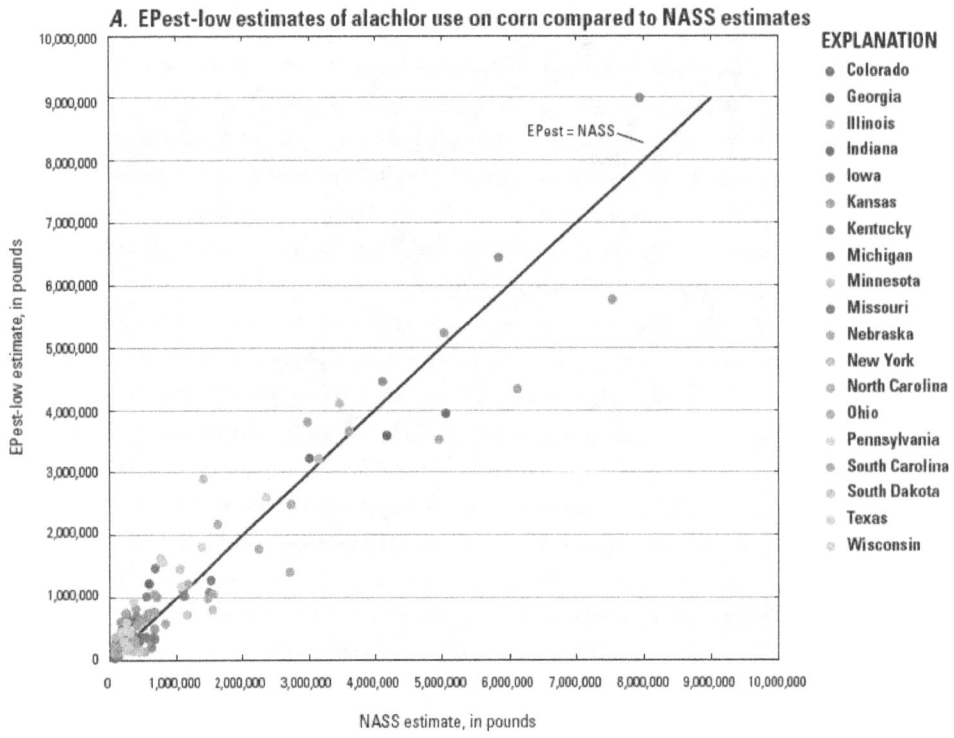

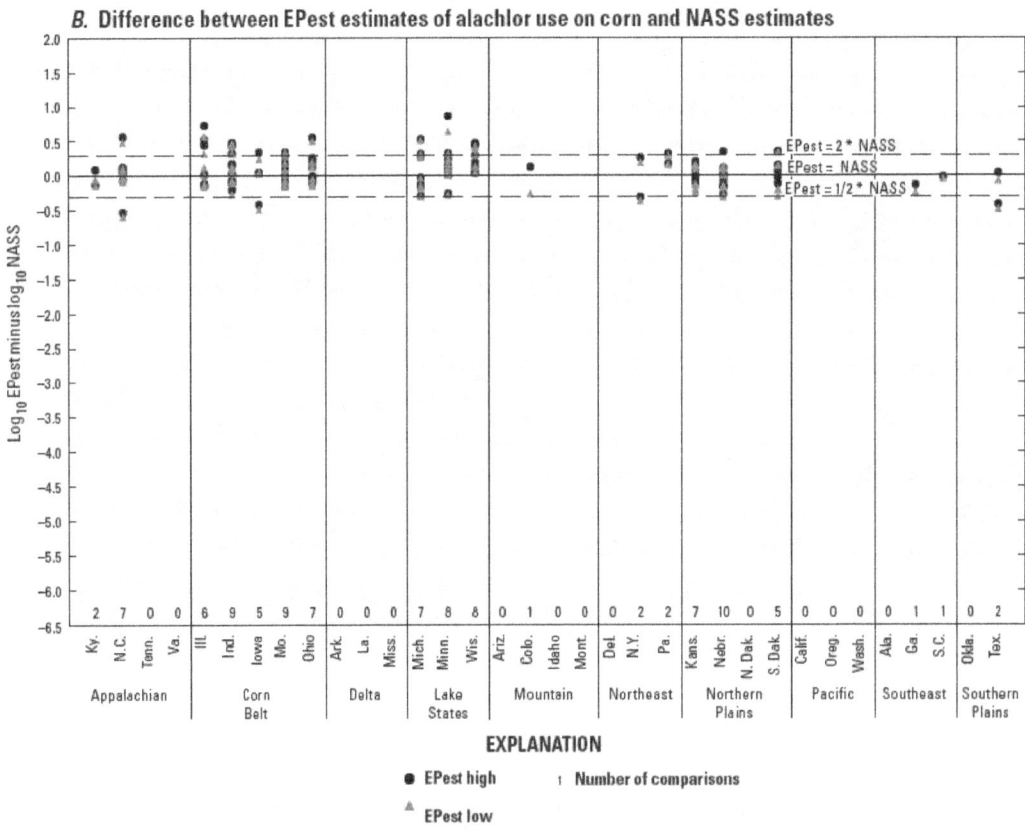

Figure 12. Comparison of EPest and National Agricultural Statistics Service (NASS) state estimates of alachlor use on corn: (*A*) EPest-low estimates compared to NASS estimates, and (*B*) Difference between EPest estimates and NASS estimates ($\log_{10}$ EPest – $\log_{10}$ NASS).

Atrazine

For various years from 1992 to 2003, 146 EPest-low and EPest-high estimates of atrazine use on corn were compared with NASS use estimates for 20 states located in the Appalachian, Corn Belt, Lake States, Mountain, Northeast, Northern Plains, Southeast, and Southern Plains regions. Both EPest-low and EPest-high estimates were significantly different than NASS use estimates ($p < 0.05$). The medians of the RE distributions comparing EPest-low and EPest-high to NASS estimates were both 7 percent greater, indicating a general tendency for EPest estimates to be slightly greater than NASS estimates. Both EPest-low and EPest-high had correlation coefficients of 0.97 with NASS use estimates, which were among the strongest correlations between pesticide use estimates in this study. The relation between EPest and NASS estimates of atrazine estimates is shown in *figure 13A*, and the differences between NASS estimates and both EPest-low and EPest-high estimates are shown by region and state in *figure 13B*.

Almost all of the EPest and NASS estimates (142 of 146) differed by less than a factor of two (*fig. 13B*), but a majority of EPest estimates were slightly greater than NASS estimates. EPest and NASS use estimates were about the same for the Appalachian, Corn Belt, Northeast, and Southeast regions, but greater differences were found for one or more estimates from the Lake States, Mountain, and Northern Plains regions.

Bentazon

For various years from 1992 through 2001, 17 EPest-low and EPest-high estimates of bentazon use on corn estimates were compared with NASS estimates for four states from the Corn Belt and Lake States regions. Both EPest-low and Epest-high estimates significantly differed from NASS use estimates ($p < 0.05$). The medians of the RE distributions comparing EPest-low and EPest-high to NASS estimates were 39 and 30 percent less, respectively, indicating a general tendency for EPest estimates to be less than NASS estimates. The correlation coefficients for the relation between the EPest and NASS estimates were 0.42 for EPest-low and 0.34 for EPest-high. The relation between EPest-low and NASS estimates of bentazon use on corn is shown in *figure 14A*, and the differences between NASS estimates and both EPest-low and EPest-high estimates are shown by region and state in *figure 14B*.

About one-half (9 of 17) of the EPest-low estimates and 65 percent (11 of 17) of the EPest-high estimates differed by less than a factor of two from NASS estimates. There were large differences between the EPest estimates and NASS use estimates for some states and years, which, in conjunction with a relatively small sample size, likely contributes to the poor correlation between the estimates.

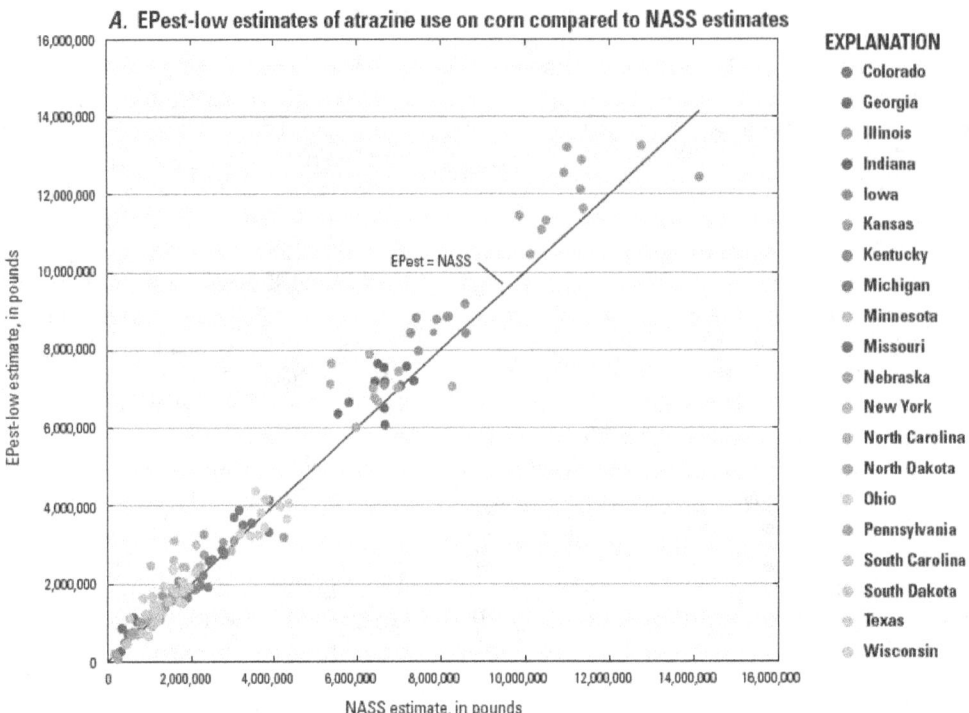

A. EPest-low estimates of atrazine use on corn compared to NASS estimates

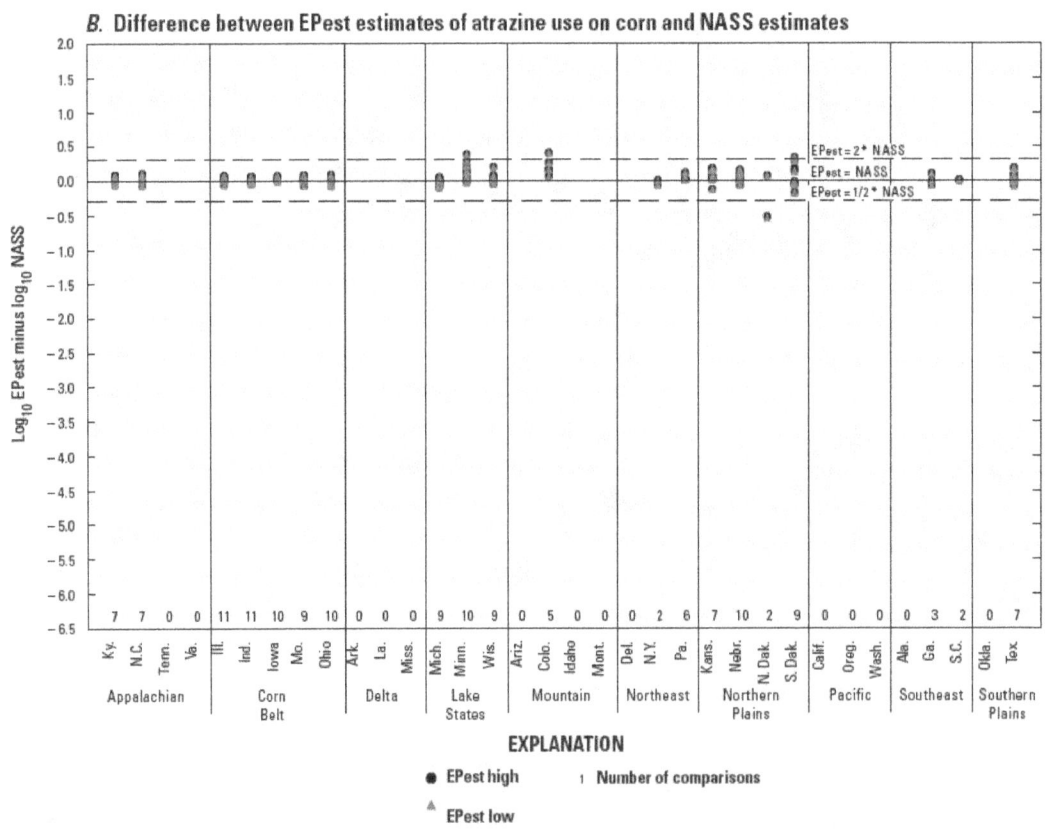

B. Difference between EPest estimates of atrazine use on corn and NASS estimates

Figure 13.   Comparison of EPest and National Agricultural Statistics Service (NASS) state estimates of atrazine use on corn: (*A*) EPest-low estimates compared to NASS estimates, and (*B*) Difference between EPest estimates and NASS estimates (log$_{10}$ EPest – log$_{10}$ NASS).

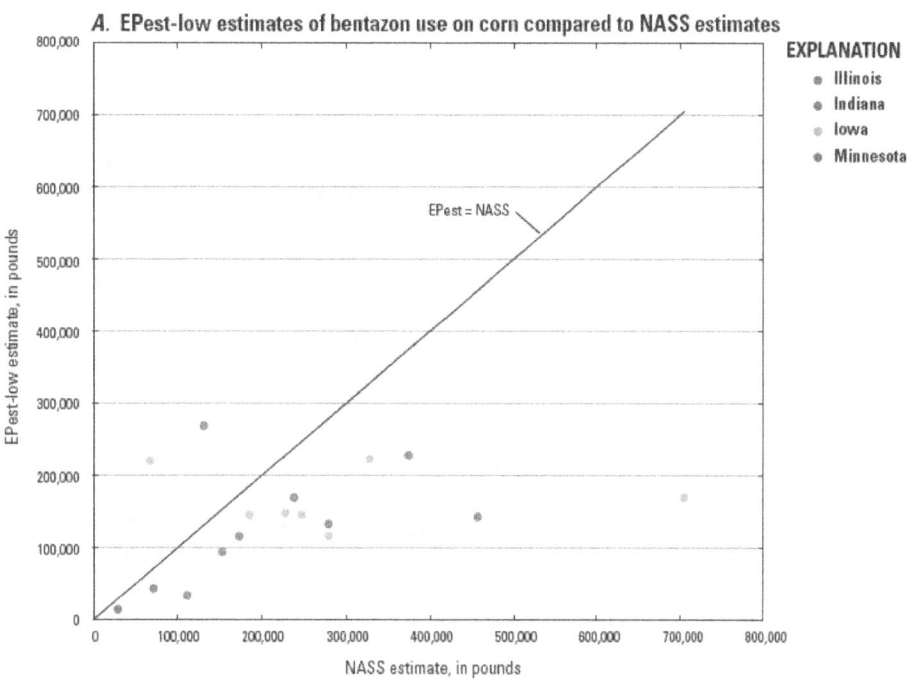

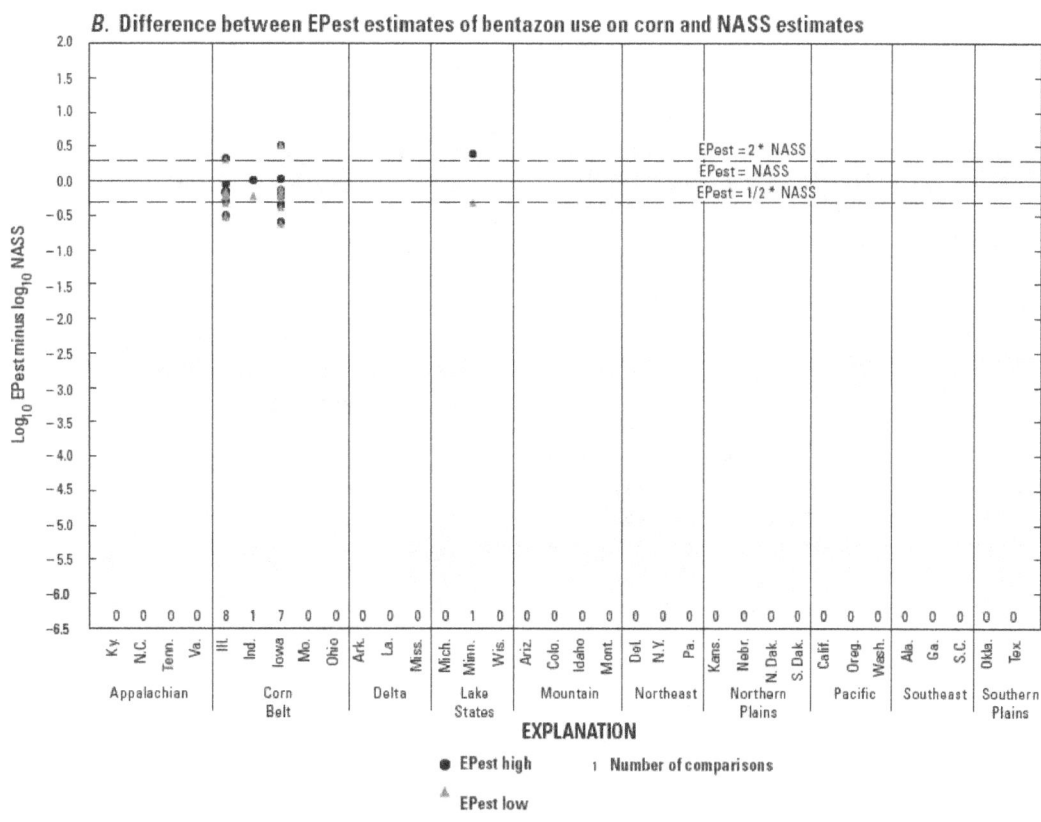

Figure 14. Comparison of EPest and National Agricultural Statistics Service (NASS) state estimates of bentazon use on corn: (*A*) EPest-low estimates compared to NASS estimates, and (*B*) Difference between EPest estimates and NASS estimates ($\log_{10}$ EPest – $\log_{10}$ NASS).

Butylate

Sixteen EPest-low and EPest-high estimates of butylate use on corn estimates were compared with NASS estimates for eight states from the Appalachian, Corn Belt, Northern Plains, and Southeast regions from 1992 through 1994. Only EPest-low estimates significantly differed from NASS use estimates (p < 0.05). The medians of the RE distributions comparing EPest-low and EPest-high to NASS estimates were 47 and 16 percent less, respectively, indicating a general tendency for EPest estimates to be less than NASS estimates. The correlation coefficient for comparison to NASS estimates to EPest-low was 0.91 and was 0.81 for EPest-high. The relation between EPest-low and NASS estimates for butylate use is shown in *figure 15A*, and the differences between NASS estimates and both EPest-low and EPest-high are shown by region and state in *figure 15B*.

The majority of EPest estimates (14 of 16 EPest-low and 10 of 16 EPest-high) were less than NASS estimates, but there was a fairly strong correlation between the estimates. Most EPest-low butylate estimates were 15 to 80 percent less than NASS estimates.

Fluometuron

For various years from 1992 through 2005, 76 EPest and NASS estimates of fluometuron use on cotton were compared for 11 states from the Appalachian, Corn Belt, Delta, Mountain, Southeast, and Southern Plains regions. Both EPest-low and EPest-high estimates significantly differed (p <0.05) from NASS estimates. The medians of the RE distributions comparing EPest-low and EPest-high to NASS estimates were 12 and 14 percent greater, respectively, indicating a general tendency for EPest estimates to be slightly greater than NASS estimates. Both EPest-low and EPest-high had correlation coefficients of 0.93 with NASS use estimates. The relation between EPest-low and NASS estimates for fluometuron is shown in *figure 16A*, and the differences between NASS estimates and both EPest-low and EPest-high rate estimates are shown by region and state in *figure 16B*.

The majority of the EPest-low (68 of 76) and EPest-high (67 of 76) estimates differed from NASS use estimates by less than a factor of two. EPest estimates tended to be greater than NASS estimates for most of the regions compared, including one or more estimates for states from the Mountain, Southeast and Southern Plains regions, which were at least twice NASS estimates. EPest totals tended to be less than NASS use estimates for some of the states in the Appalachian, Delta, and Southern Plains, however.

Glyphosate

EPest and NASS estimates of glyphosate use were compared for corn, cotton, soybeans, spring wheat, and winter wheat crops. EPest estimates significantly differed from NASS estimates for the crops evaluated, except for soybeans, which also had the highest correlation coefficient between EPest and NASS estimates and the lowest median RE. Comparisons of EPest and NASS estimates for glyphosate use on spring and winter wheat crops showed low correlation coefficients and small sample sizes, which limits the power of the statistical tests. EPest and NASS estimates of glyphosate use on corn and cotton are discussed in the following sections.

Corn

For glyphosate use on corn, 121 EPest and NASS estimates were compared from 19 states from the Appalachian, Corn Belt, Lake States, Mountain, Northeast, Northern Plains, Southeast, and Southern Plains regions. Both EPest-low and EPest-high estimates significantly differed (p <0.05) from NASS estimates. The medians of the RE distributions comparing EPest-low and EPest-high to NASS estimates were both 34 percent greater, indicating a general tendency for EPest estimates to be greater than NASS estimates. Correlation coefficients for EPest-low and NASS comparisons were 0.78 and were 0.79 for EPest-high. The relation between EPest-low and NASS estimates for glyphosate use on corn is shown in *figure 17A*, and the differences between NASS estimates and both EPest-low and EPest-high are shown by region and state in *figure 17B*.

Most of the EPest and NASS estimates (90 or more of 121) differed by less than a factor of two. EPest-low and EPest-high estimates for the Corn Belt, Lake States, Northeast, Southeast, and Southern Plains regions tended to be greater than NASS estimates, and estimates for one or more states in each of these regions had EPest estimates that were more than twice the NASS estimate (*fig. 17B*).

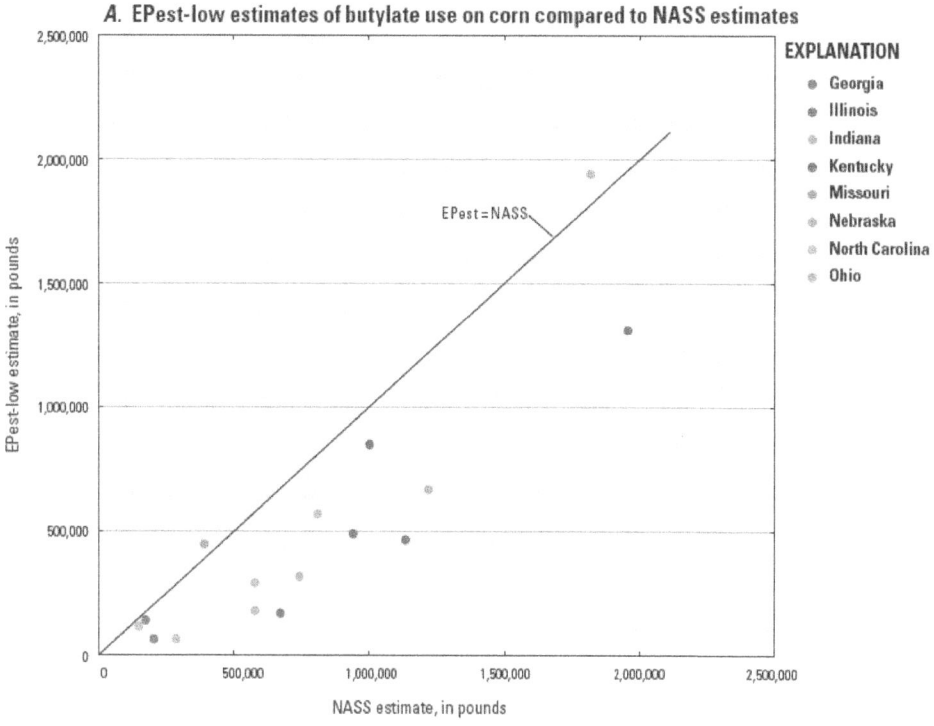

A. EPest-low estimates of butylate use on corn compared to NASS estimates

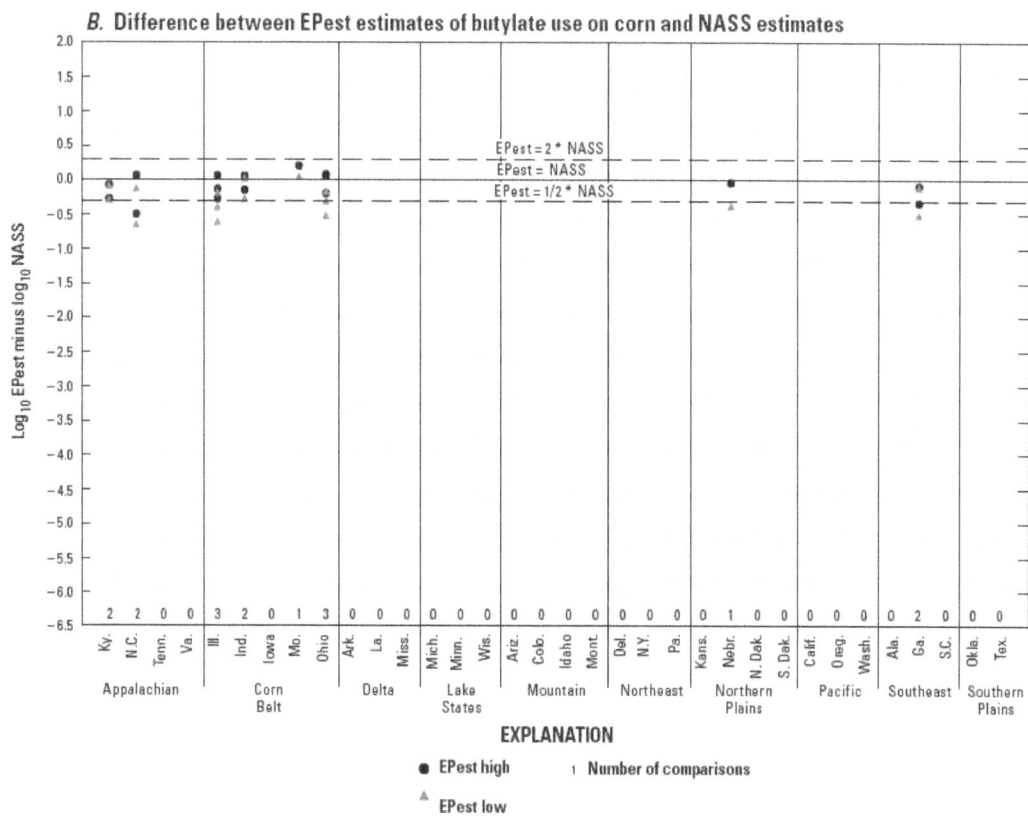

B. Difference between EPest estimates of butylate use on corn and NASS estimates

Figure 15. Comparison of EPest and National Agricultural Statistics Service (NASS) state estimates of butylate use on corn: (*A*) EPest-low estimates compared to NASS estimates, and (*B*) Difference between EPest estimates and NASS estimates ($\log_{10}$ EPest – $\log_{10}$ NASS).

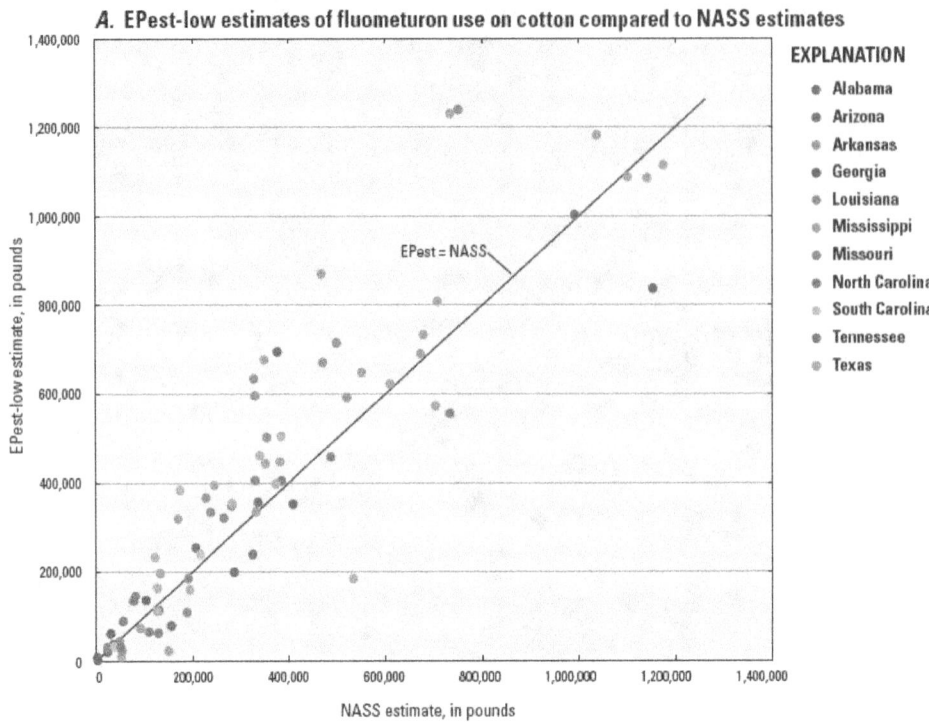

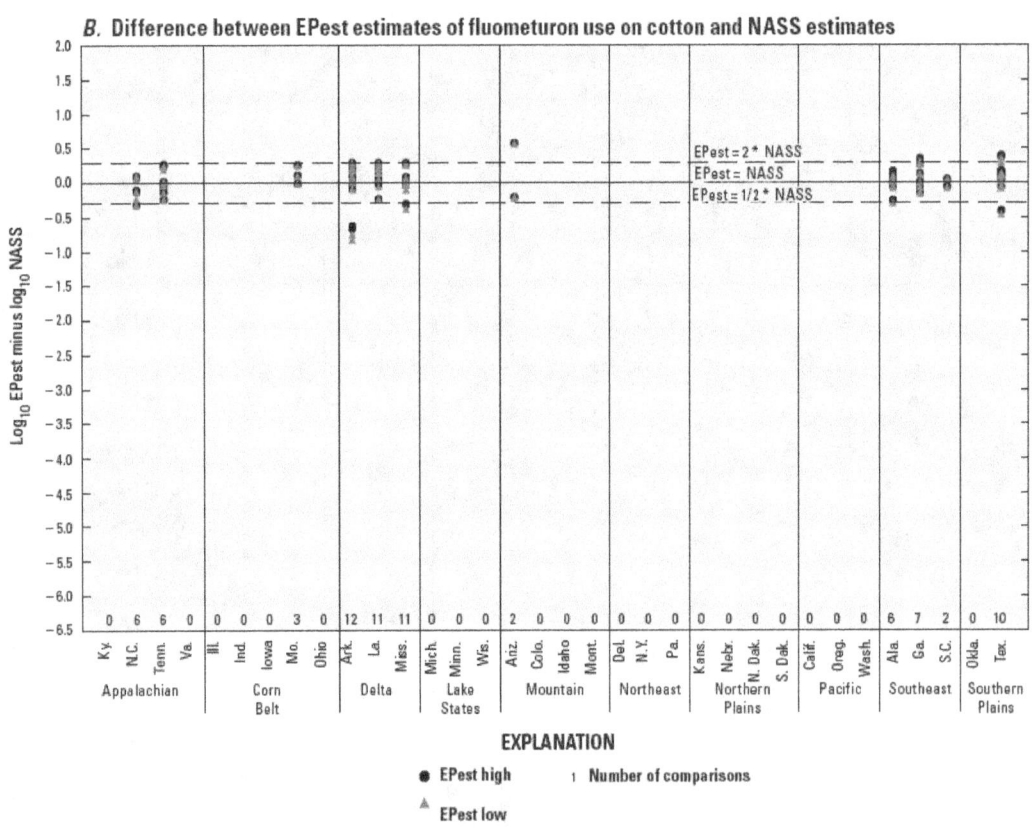

Figure 16. Comparison of EPest and National Agricultural Statistics Service (NASS) state estimates of fluometuron use on cotton: (*A*) EPest-low estimates compared to NASS estimates, and (*B*) Difference between EPest estimates and NASS estimates ($\log_{10}$ EPest – $\log_{10}$ NASS).

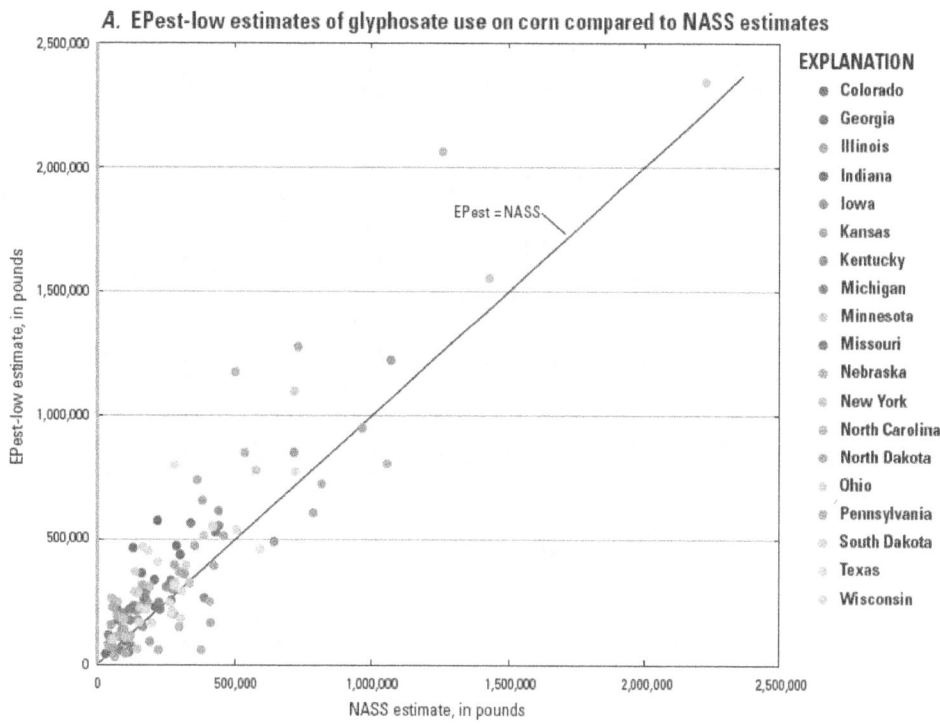

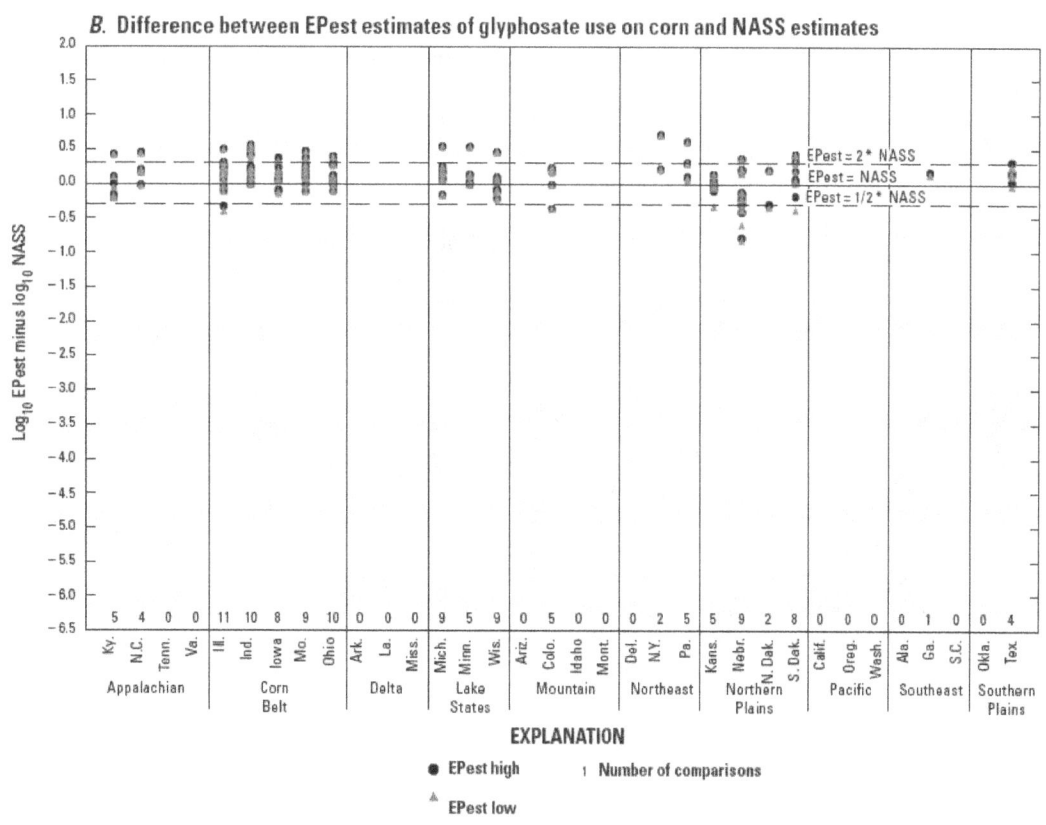

Figure 17. Comparison of EPest and National Agricultural Statistics Service (NASS) state estimates of glyphosate use on corn: (*A*) EPest-low estimates compared to NASS estimates, and (*B*) Difference between EPest estimates and NASS estimates ($\log_{10}$ EPest – $\log_{10}$ NASS).

Cotton

For various years from 1992 through 2005, 83 EPest-low and EPest-high estimates of glyphosate use on cotton were compared with NASS estimates for 12 states from Appalachian, Corn Belt, Delta, Mountain, Pacific, Southeast, and Southern Plains regions. Both EPest-low and EPest-high estimates significantly differed (p <0.05) from NASS use estimates. The medians of the RE distributions comparing EPest-low and EPest-high to NASS estimates were both 30 percent greater, indicating a general tendency for EPest estimates to be greater than NASS estimates. Correlation coefficients for EPest-low and NASS comparisons were 0.93 and were 0.92 for EPest-high. The relation between EPest and NASS estimates of glyphosate use on cotton is shown in *figure 18A*, and the differences between NASS estimates and both EPest-low and EPest-high estimates are shown by region and state in *figure 18B*.

Most EPest and NASS estimates (63 of 83) differed by less than a factor of two. EPest estimates for the Appalachian, Delta and Corn Belt regions bracketed NASS use estimates, whereas in most other regions, EPest estimates were greater than NASS use estimates. One reason for this difference could be that EPest pesticide totals include pesticide use on both upland and Pima cotton, whereas NASS reports pesticide use for upland cotton only.

Metolachlor

Corn

For various years from 1992 through 2003, 130 EPest-low and EPest-high estimates of metolachlor use on corn were compared with NASS estimates for 18 states from the Appalachian, Corn Belt, Lake States, Mountain, Northeast, Southeast, and Northern and Southern Plains regions. Both EPest-low and EPest-high estimates significantly differed (p <0.05) from NASS use estimates. The medians of the RE distributions comparing EPest-low and EPest-high to NASS estimates were 10 and 7 percent lower, respectively, indicating a general tendency for EPest estimates to be less than NASS estimates. Correlation coefficients for EPest-low and NASS comparisons were 0.87 and were 0.88 for EPest-high. The relation between EPest-low and NASS estimates of metolachlor use on corn is shown in *figure 19A*, and the differences between NASS estimates and both EPest-low and EPest-high estimates are shown by region and state in *figure 19B*.

Most EPest estimates differed from NASS estimates by less than a factor of two, and estimates for most states bracketed NASS estimates. From 1998 through 2003, however, there were 30 EPest-low and EPest-high estimates that were more than 50 percent lower than NASS estimates, representing some of the greatest underestimates of EPest compared to NASS. Beginning in the late 1990s and early 2000s, metolachlor use was being replaced by use of *S*-metolachlor. It is possible that this difference in estimates could be related to how metolachlor and *S*-metolachlor were surveyed and reported. NASS estimates for metolachlor may have also included information for the related compound *S*-metolachlor. For example, beginning in 2002, EPest-low estimates of metolachlor use were zero for several states, such as Illinois and Iowa, while NASS reported several hundred pounds to over one million pounds of metolachlor use in these same states.

Soybeans

For various years from 1992 through 2000, 89 EPest-low and EPest-high estimates of metolachlor use on soybeans were compared with NASS estimates for 18 states from the Appalachian, Corn Belt, Delta, Lake States, Northeast, and Northern Plains regions. Only EPest-high estimates significantly differed (p <0.05) from NASS estimates. The medians of the RE distributions comparing EPest-low and EPest-high to NASS estimates were 8 and 19 percent greater, respectively, indicating a general tendency for EPest estimates to be greater than NASS estimates. The correlation coefficients for EPest-low and NASS comparisons were 0.76 and were 0.75 for EPest-high. The relation between EPest-low and NASS estimates of metolachlor use on soybeans are shown in *figure 20A*, and the differences between NASS estimates and both EPest-low and EPest-high estimates are shown by region and state in *figure 20B*.

The majority (71 of 89) of EPest and NASS estimates differed by less than a factor of two (*fig. 20B*). EPest estimates for most regions tended to be greater than NASS estimates, but in the Appalachian region, they tended to be less than NASS estimates.

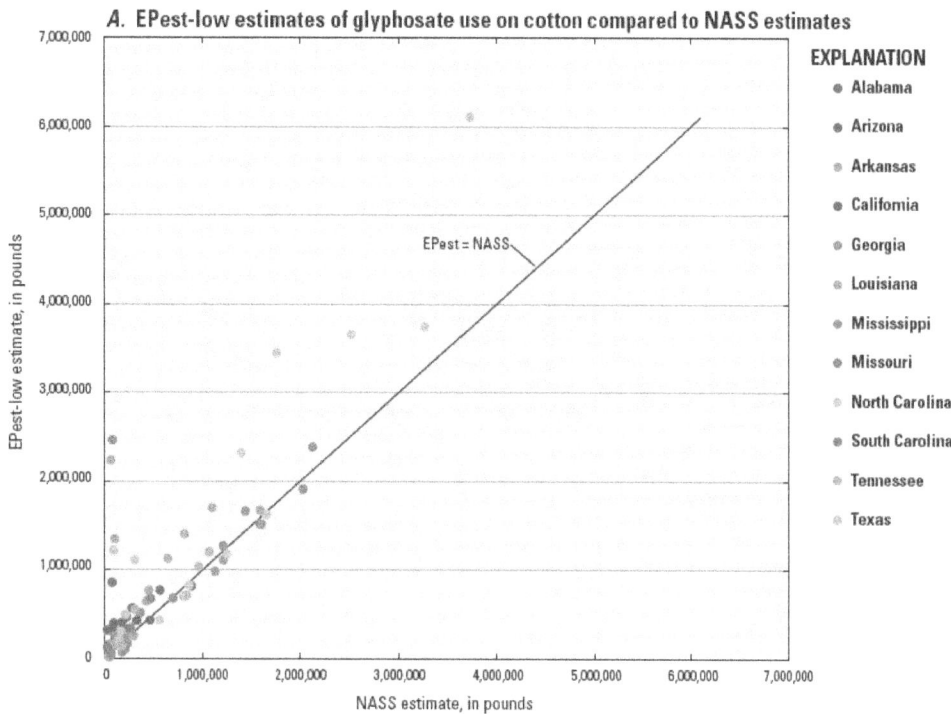

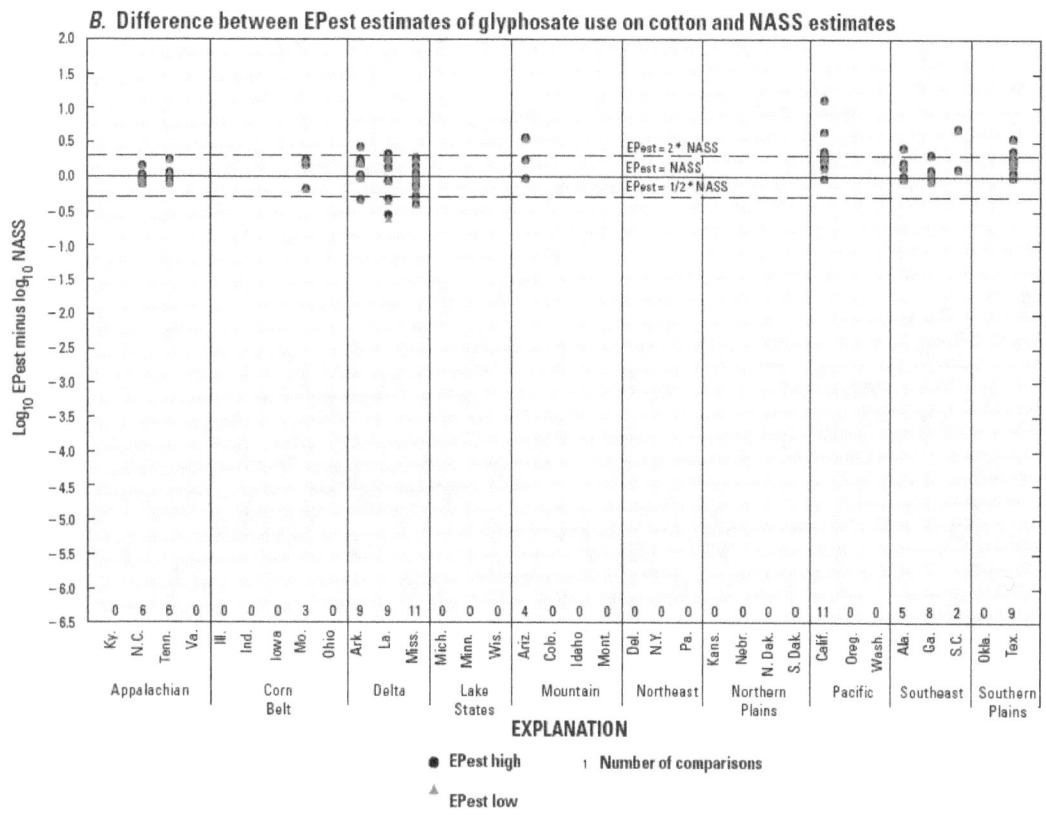

Figure 18. Comparison of EPest and National Agricultural Statistics Service (NASS) state estimates of glyphosate use on cotton: (*A*) EPest-low estimates compared to NASS estimates, and (*B*) Difference between EPest estimates and NASS estimates ($\log_{10}$ EPest – $\log_{10}$ NASS).

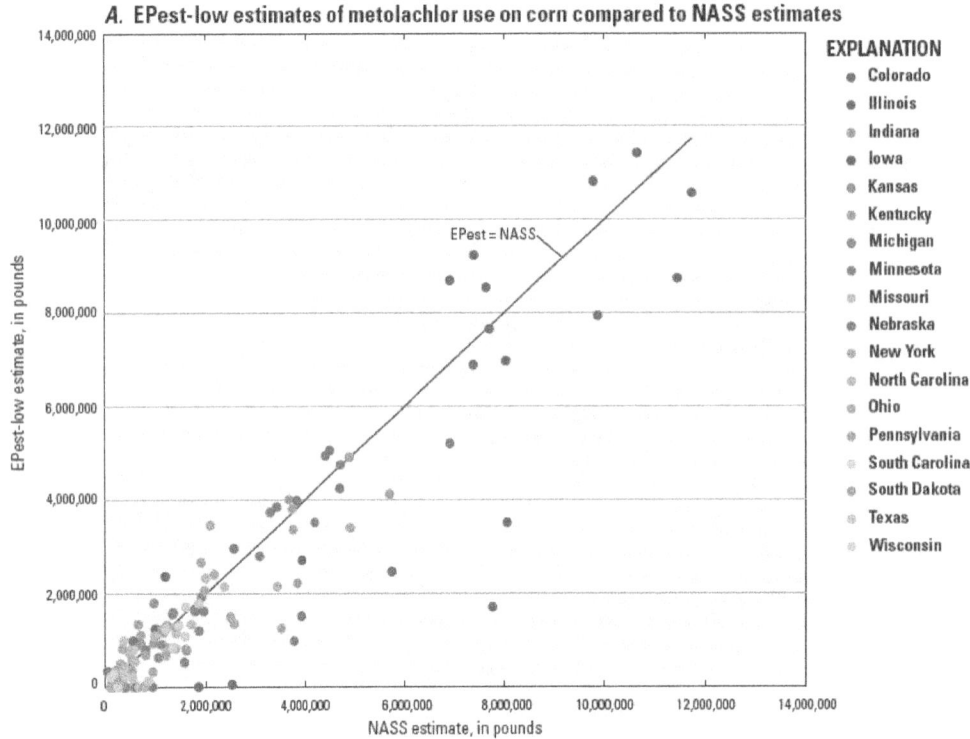

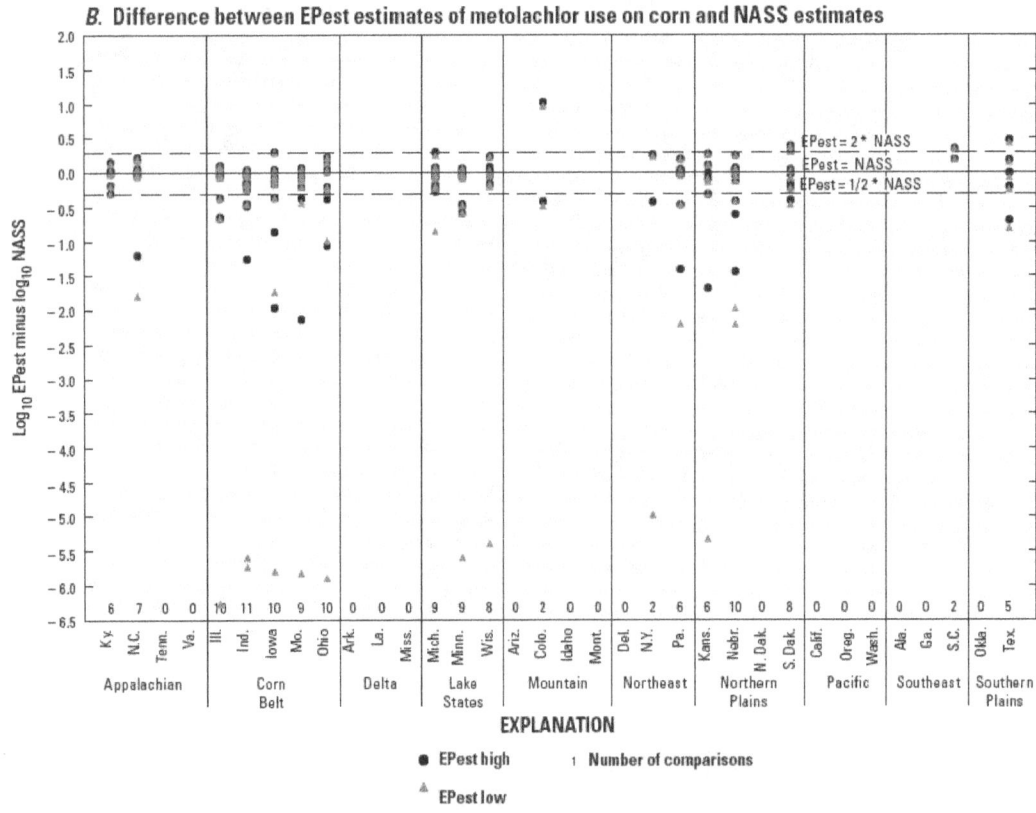

Figure 19.   Comparison of EPest and National Agricultural Statistics Service (NASS) state estimates of metolachlor use on corn:
(*A*) EPest-low estimates compared to NASS estimates, and (*B*) Difference between EPest estimates and NASS estimates
($\log_{10}$ EPest – $\log_{10}$ NASS).

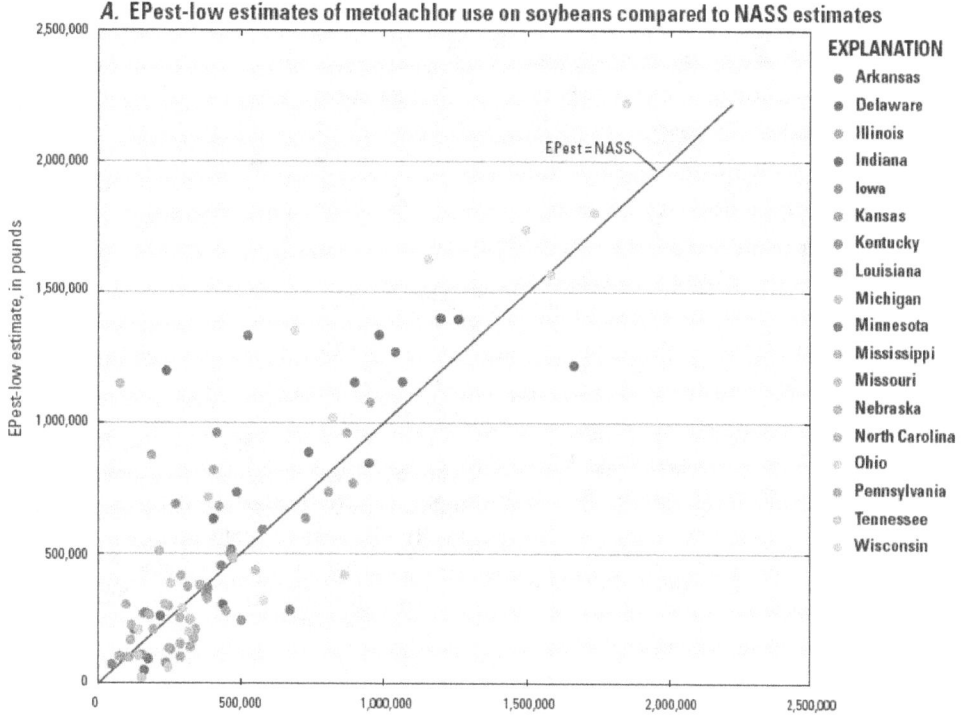

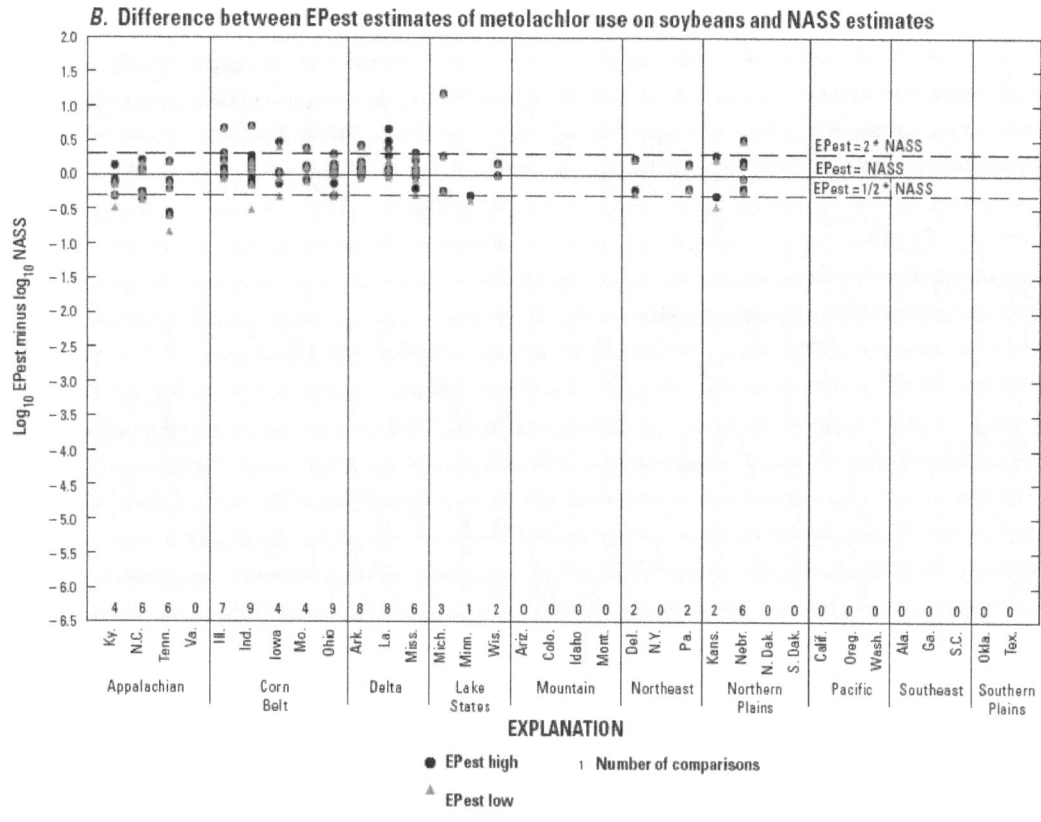

Figure 20.   Comparison of EPest and National Agricultural Statistics Service (NASS) state estimates of metolachlor use on soybeans:
(*A*) EPest-low estimates compared to NASS estimates, and (*B*) Difference between EPest estimates and NASS estimates
($\log_{10}$ EPest – $\log_{10}$ NASS).

Metribuzin

For various years from 1992 through 2006, 108 EPest-low and Epest-high estimates of metribuzin use on soybeans were compared with NASS estimates in 19 states located in the Appalachian, Corn Belt, Delta, Lake States, Northeast, and Southeast regions. Only EPest-high estimates were significantly different (p <0.05) from NASS estimates. The medians of the RE distributions comparing EPest-low and EPest-high to NASS estimates were 12 and 18 percent greater, respectively, indicating a general tendency for EPest estimates to be slightly greater than NASS estimates. Correlation coefficients for EPest-low and NASS comparisons were 0.81 and were 0.80 for EPest-high. The relation between EPest-low and NASS estimates of metribuzin use is shown in *figure 21A*, and the differences between NASS estimates and both EPest-low and EPest-high estimates are shown by region and state in *figure 21B*.

The majority of EPest estimates were within a factor of two of NASS estimates (*fig. 21B*). EPest estimates for all of the regions bracketed NASS estimates, but estimates from Arkansas and Nebraska showed some large differences.

Nicosulfuron

For various years from 1992 through 2003, 127 EPest-low and EPest-high estimates of nicosulfuron use on corn were compared with NASS estimates for 20 states located in Appalachian, Corn Belt, Lake States, Mountain, Northeast, Northern Plains, Southeast, and Southern Plains regions. EPest-low and EPest-high estimates significantly differed (p <0.05) from NASS estimates. The medians of the RE distributions comparing EPest-low and EPest-high to NASS estimates were 14 and 17 percent greater, respectively, indicating a general tendency for EPest estimates to be greater than NASS estimates Correlation coefficients for EPest and NASS comparisons were 0.84 for both EPest-low and EPest-high. The relation between EPest-low and NASS estimates of nicosulfuron use on corn is shown in *figure 22A*, and the differences between NASS estimates and both EPest-low and EPest-high estimates are shown by region and state in *figure 22B*.

Most of the EPest estimates were greater than NASS estimates, and the majority (98 of 127) of comparisons differed by less than a factor of two, although one or more EPest estimates from the Appalachian, Corn Belt, Lake States, Northeast, Northern Plains, and Southern Plains regions were at least twice NASS estimates. For some of the same states in these regions, however, EPest totals were half or less of NASS estimates.

S-Metolachlor

For 17 states from the Appalachian, Corn Belt, Lake States, Mountain, Northeast, Northern, and Southern Plains regions from 2001 through 2003, 39 EPest-low and EPest-high estimates of S-metolachlor use on corn were compared with NASS estimates. Only EPest-high estimates significantly differed (p <0.05) from NASS estimates. The medians of the RE distributions comparing EPest-low and EPest-high to NASS estimates were 8 and 16 percent greater, respectively, indicating a general tendency for EPest estimates to be slightly greater than NASS estimates. Correlation coefficients for EPest-low and NASS comparisons were 0.90 and were 0.91 for EPest-high. The relation between EPest and NASS estimates of S-metolachlor use is shown in *figure 23A*, and the differences between NASS estimates and both EPest-low and EPest-high estimates are shown by region and state in *figure 23B*.

EPest and NASS estimates for the majority (36 of 39) of states and years were within a factor of two (*fig. 23B*). EPest estimates for the Corn Belt, Mountain, Northern Plains, and Southern Plains regions tended to be greater than NASS estimates, whereas EPest estimates for the Lake States and Northeast tended to be less than NASS estimates.

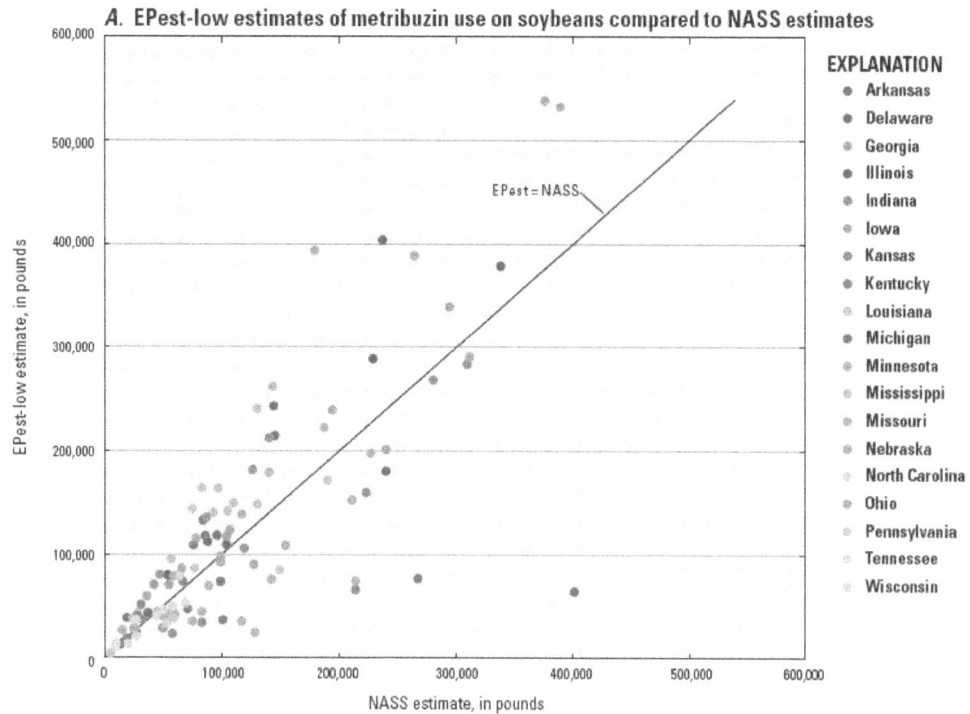

A. EPest-low estimates of metribuzin use on soybeans compared to NASS estimates

EXPLANATION
- Arkansas
- Delaware
- Georgia
- Illinois
- Indiana
- Iowa
- Kansas
- Kentucky
- Louisiana
- Michigan
- Minnesota
- Mississippi
- Missouri
- Nebraska
- North Carolina
- Ohio
- Pennsylvania
- Tennessee
- Wisconsin

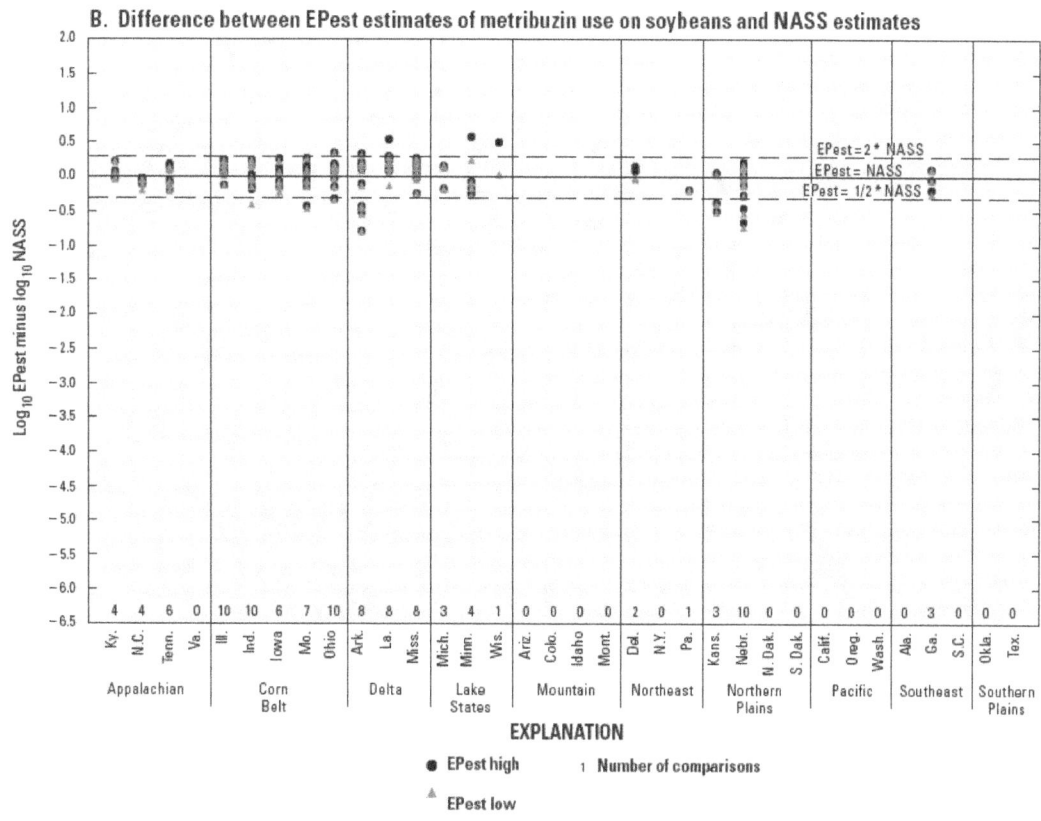

B. Difference between EPest estimates of metribuzin use on soybeans and NASS estimates

EXPLANATION
- EPest high   1   Number of comparisons
- EPest low

Figure 21. Comparison of EPest and National Agricultural Statistics Service (NASS) state estimates of metribuzin use on soybeans: (*A*) EPest-low estimates compared to NASS estimates, and (*B*) Difference between EPest estimates and NASS estimates ($\log_{10}$ EPest – $\log_{10}$ NASS).

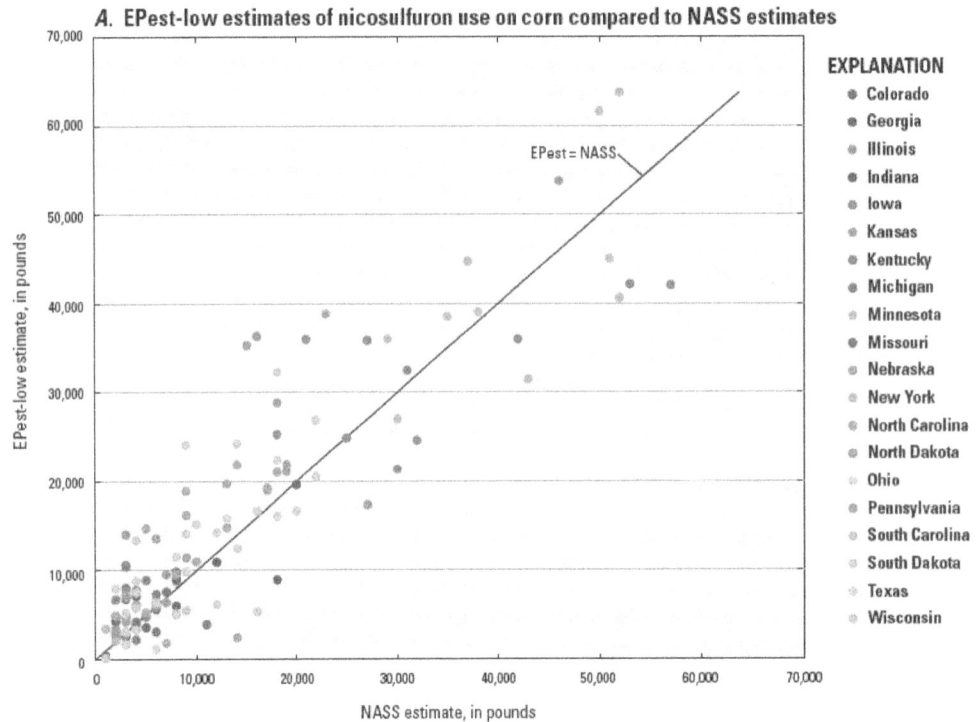

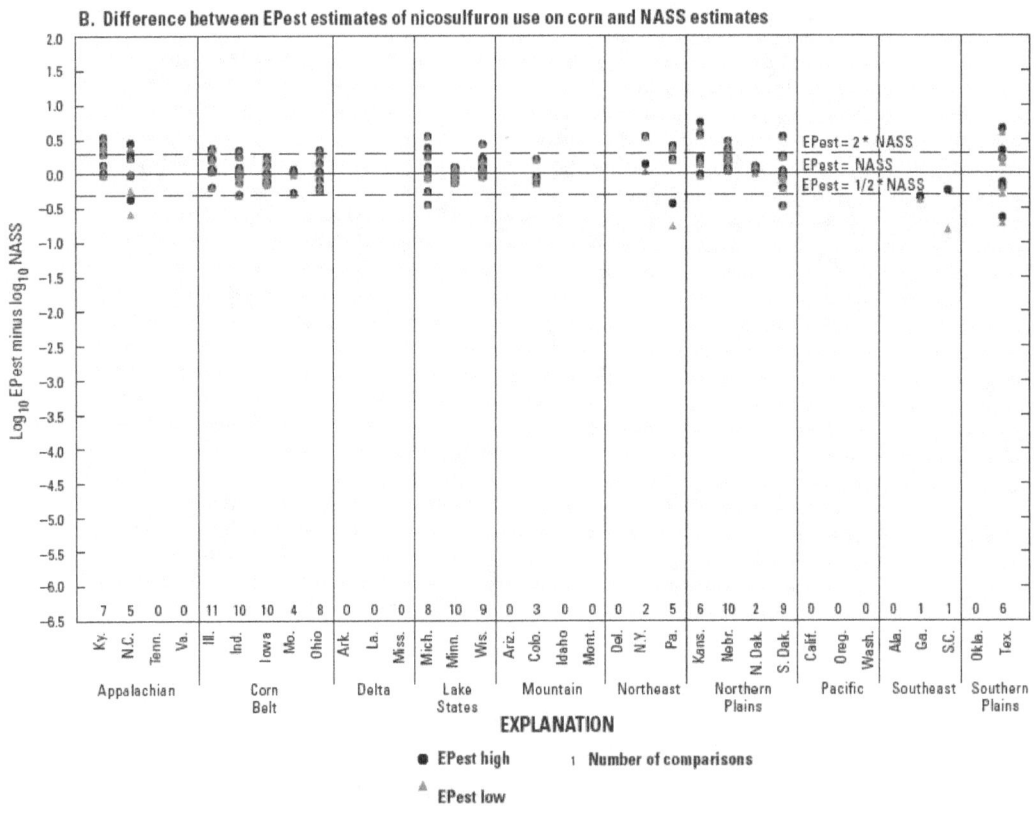

Figure 22.   Comparison of EPest and National Agricultural Statistics Service (NASS) state estimates of nicosulfuron use on corn:
(*A*) EPest-low estimates compared to NASS estimates, and (*B*) Difference between EPest estimates and NASS estimates
($\log_{10}$ EPest – $\log_{10}$ NASS).

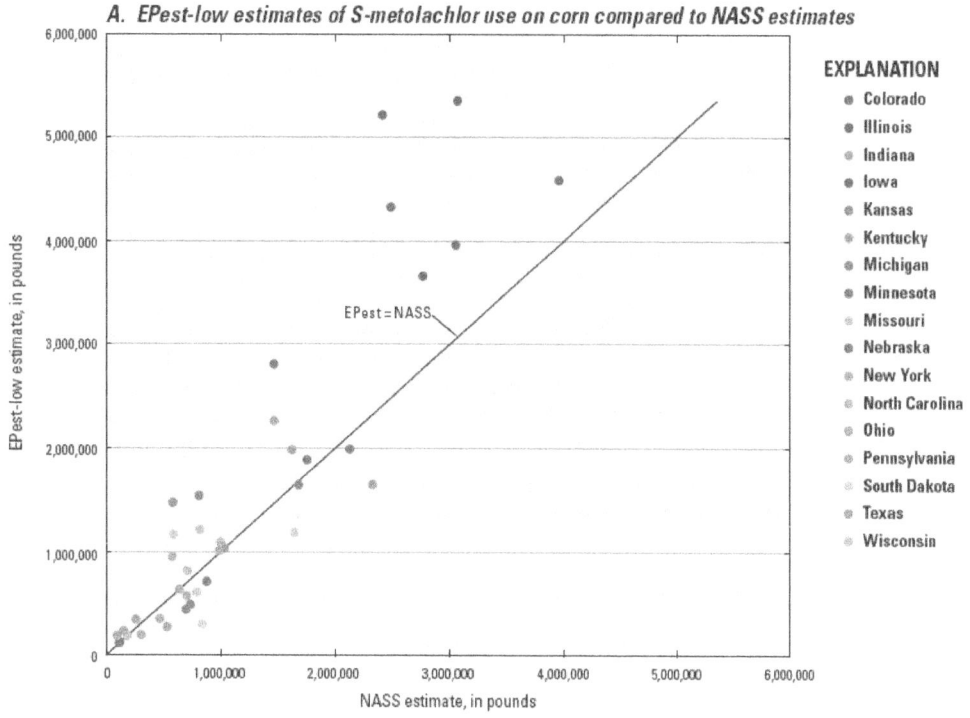

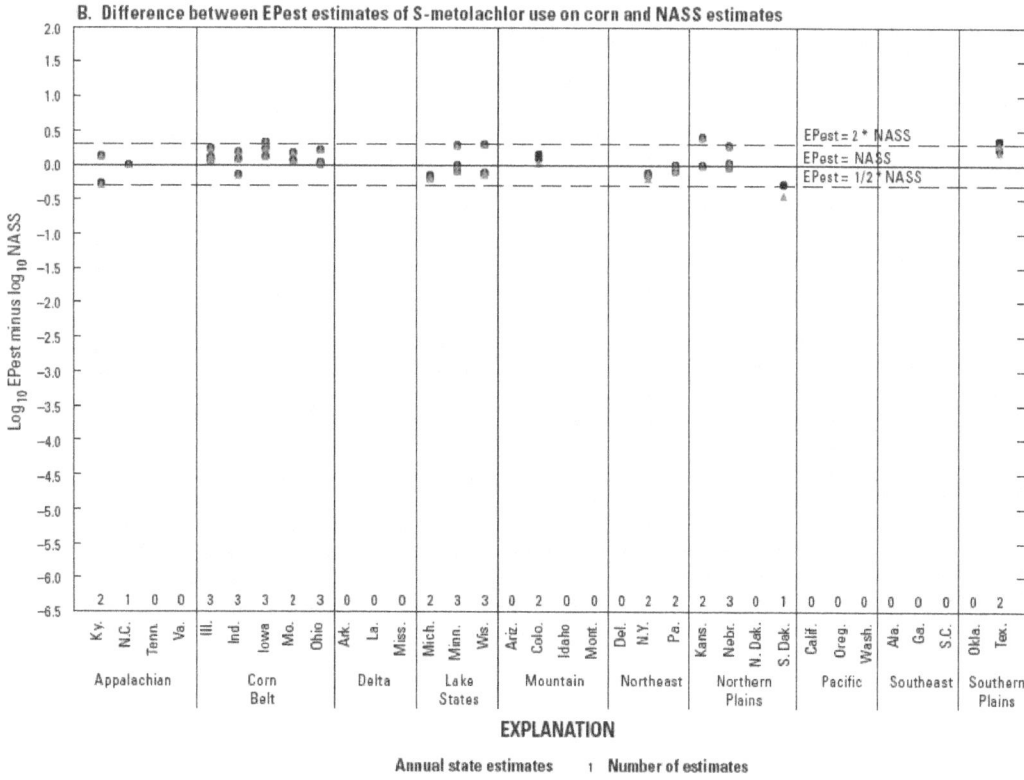

Figure 23. Comparison of EPest and National Agricultural Statistics Service (NASS) state estimates of *S*-metolachlor use on corn: (*A*) EPest-low estimates compared to NASS estimates, and (*B*) Difference between EPest estimates and NASS estimates ($\log_{10}$ EPest – $\log_{10}$ NASS).

Trifluralin

Cotton

For various years from 1992 through 2005, 90 EPest-low and EPest-high estimates of trifluralin use on cotton were compared with NASS estimates for 12 states from the Appalachian, Corn Belt, Mountain, Pacific, Southeast, and Southern Plains regions. Only EPest-high estimates significantly differed ($p < 0.05$) from NASS estimates. The medians of the RE distributions comparing EPest-low and EPest-high to NASS estimates were 6 and 10 percent greater, respectively, indicating a general tendency for EPest estimates to be slightly greater than NASS estimates. Correlation coefficients for EPest and NASS comparisons were 0.95 for both EPest-low and EPest-high. The relation between EPest-low and NASS estimates of trifluralin use on cotton is shown in *figure 24A*, and the differences between NASS estimates and both EPest-low and EPest-high estimates are shown by region and state in *figure 24B*.

The majority of EPest estimates differed from NASS estimates by less than a factor of two. The EPest estimates for most of the states in a particular region were evenly distributed around NASS use estimates. The strong correlation between estimates was driven by use estimates in Texas, which showed the least differences between EPest and NASS estimates of all the states.

Soybeans

For various years from 1992 through 2006, 97 EPest-low and EPest-high estimates of trifluralin use on soybeans were compared for 18 states from the Appalachian, Corn Belt, Delta, Lake States, Northeast, Southeast, and Northern Plains regions. Only EPest-high estimates significantly differed ($p < 0.05$) from NASS estimates. The medians of the RE distributions comparing EPest-low and EPest-high to NASS estimates were 3 and 7 percent greater, respectively, indicating a general tendency for EPest estimates to be slightly greater than NASS estimates. Correlation coefficients for EPest and NASS comparisons were 0.91 for both EPest-low and EPest-high. The relation between EPest-low and NASS estimates of trifluralin use on soybeans is shown in *figure 25A*, and the differences between NASS estimattes and both EPest-low and EPest-high estimates are shown by region and state in *figure 25B*.

The majority of EPest and NASS estimates were within a factor of two. One or more EPest and NASS estimates from every region except the Northern Plains differed by more than a factor of two. Iowa had greater trifluralin use on soybeans than other states.

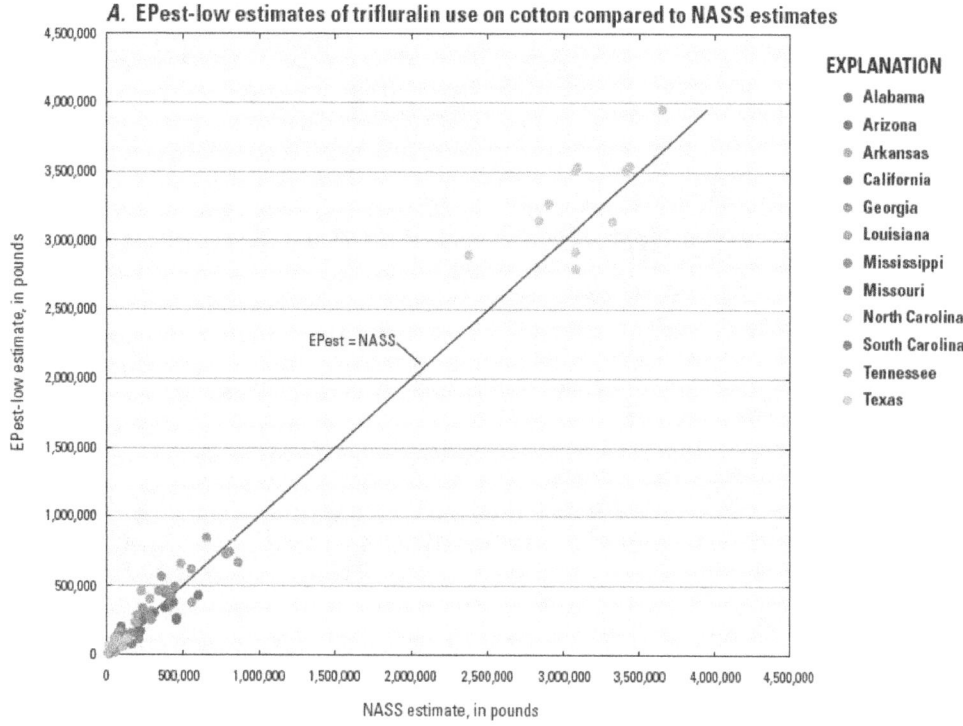

A. EPest-low estimates of trifluralin use on cotton compared to NASS estimates

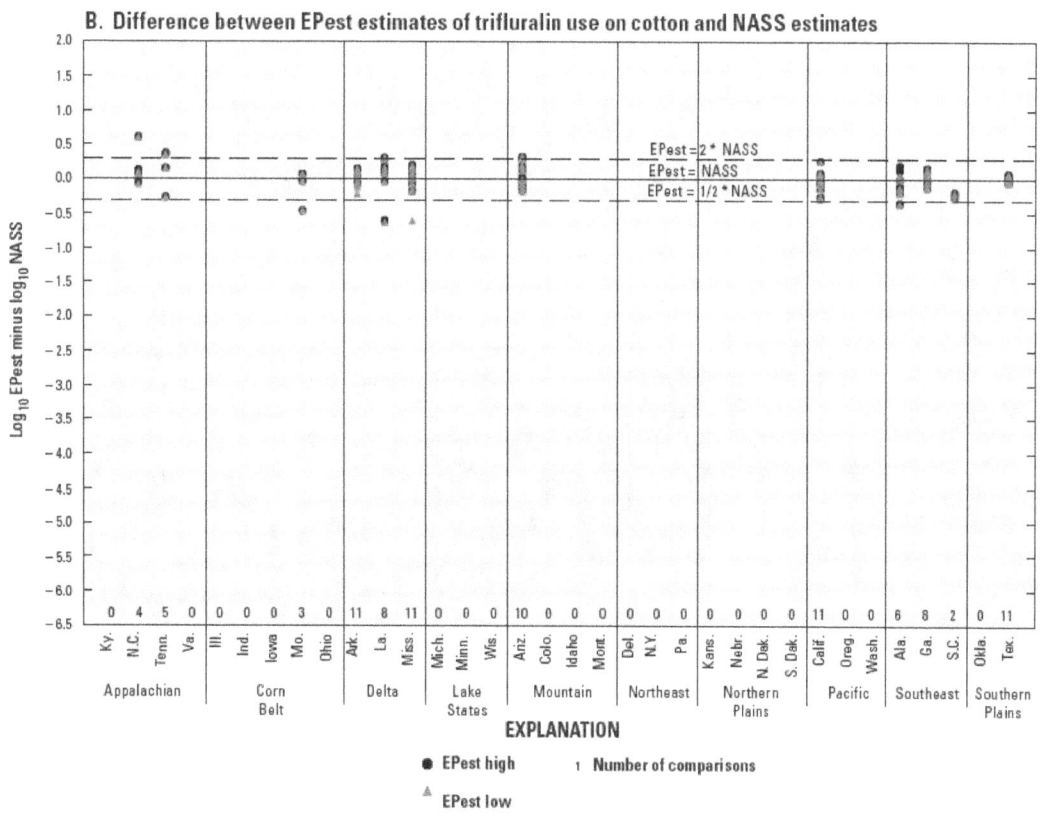

B. Difference between EPest estimates of trifluralin use on cotton and NASS estimates

Figure 24. Comparison of EPest and National Agricultural Statistics Service (NASS) state estimates of trifluralin use on cotton: (*A*) EPest-low estimates compared to NASS estimates, and (*B*) Difference between EPest estimates and NASS estimates ($\log_{10}$ EPest – $\log_{10}$ NASS).

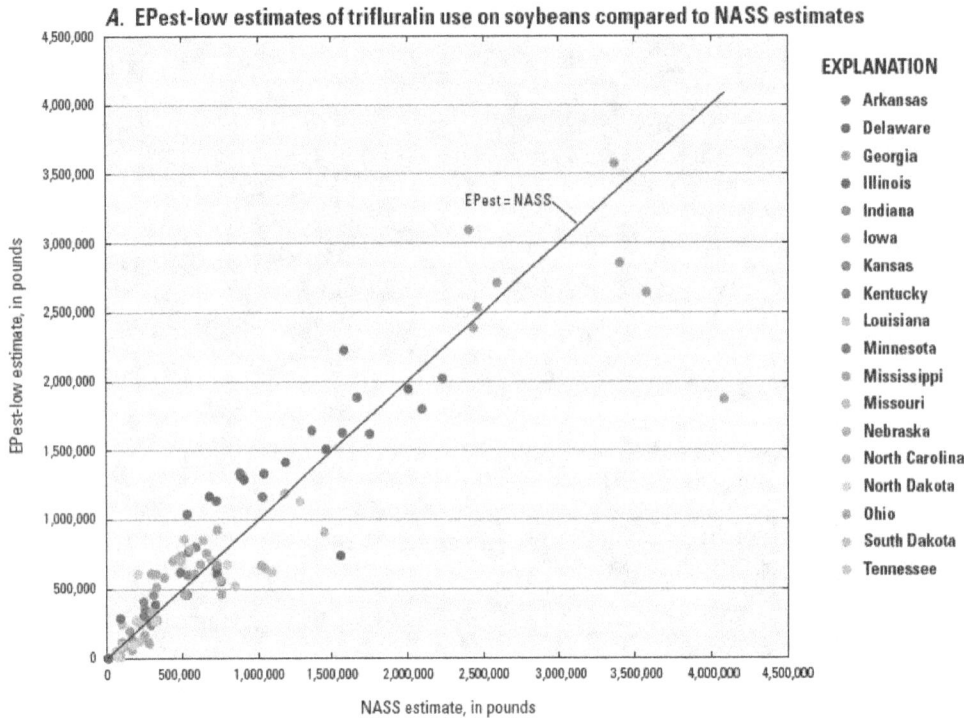

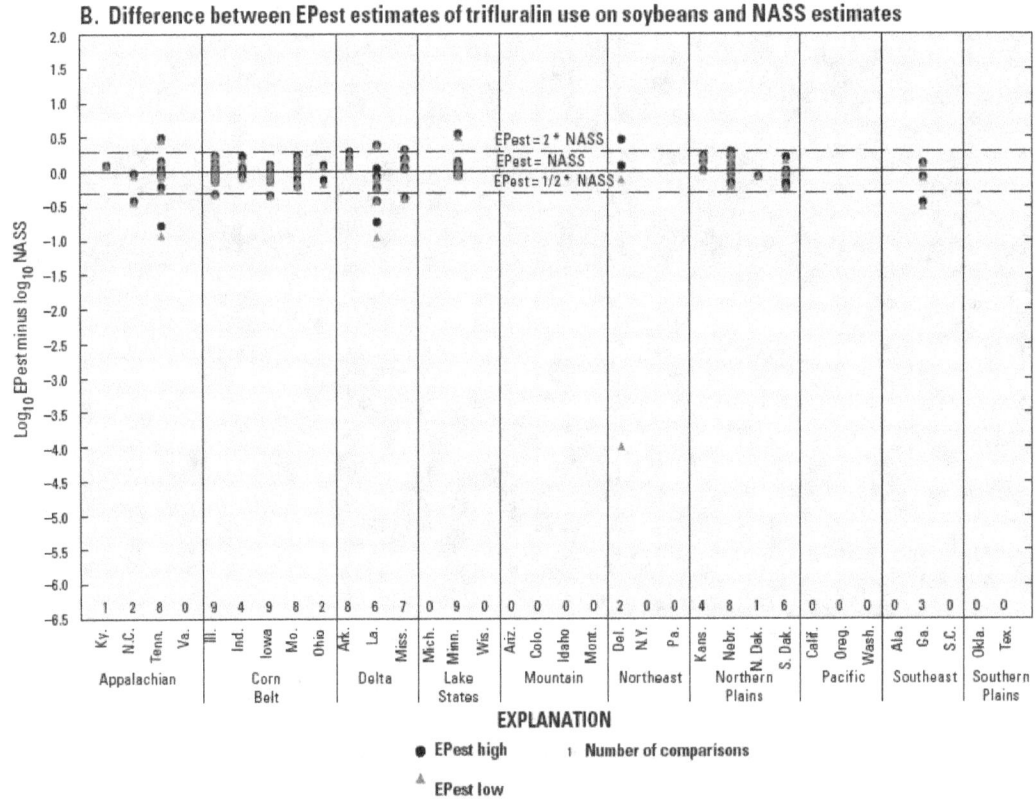

Figure 25.   Comparison of EPest and National Agricultural Statistics Service (NASS) state estimates of trifluralin use on soybeans: (*A*) EPest-low estimates compared to NASS estimates, and (*B*) Difference between EPest estimates and NASS estimates ($\log_{10}$ EPest – $\log_{10}$ NASS).

Insecticide Estimate Comparisons between EPest and NASS

EPest and NASS estimates were compared for seven insecticides used on corn, cotton, or both, as summarized in *tables 3* and *4*. Only 2 of the 10 insecticide comparisons had sample numbers greater than 50; both of these were not significant and had RE values of 0.1 or less, indicating agreement between the estimates. Most of the other comparisons were not significant and had RE values of 0.15 or less, but methomyl and methyl parathion estimates for cotton significantly differed and had RE values greater than 0.6, which are discussed in the following sections.

Methomyl

For various years from 1992 through 2003, 27 EPest-low and EPest-high estimates of methomyl use on cotton were compared with NASS estimates for 9 states from the Appalachian, Delta, Mountain, Pacific, Southeast, and Southern Plains regions. Only EPest-low estimates significantly differed ($p < 0.05$) from NASS estimates. The medians of the RE distributions comparing EPest-low and EPest-high to NASS estimates were 61 and 46 percent less, respectively, indicating a general tendency for EPest estimates to be less than NASS estimates. Correlation coefficients for EPest-low and NASS comparisons were 0.76 and were 0.74 for EPest-high. The relation between EPest-low and NASS estimates of methomyl use on cotton is shown in *figure 26A*, and the differences between NASS estimates and both EPest-low and EPest-high estimates are shown by region and state in *figure 26B*.

More than half of the EPest estimates were less than 50 percent of NASS estimates, although one EPest estimate from Arkansas was more than double the NASS estimate. The few EPest and NASS estimates for California, Georgia, and Texas were in closer agreement than the estimates for other states.

Methyl Parathion

For various years from 1992 through 2005, 50 EPest-low and EPest-high estimates of methyl parathion use on cotton were compared with NASS estimates for 8 states from the Appalachian, Corn Belt, Delta, Mountain, Southeast, and Southern Plains regions. Both EPest-low and EPest-high estimates significantly differed ($p < 0.05$) from NASS estimates. The medians of the RE distributions comparing EPest-low and EPest-high to NASS estimates were 78 and 69 percent less, respectively, indicating a general tendency for EPest estimates to be less than NASS estimates. Correlation coefficients for EPest-low and NASS comparisons were 0.47 and were 0.52 for EPest-high. The relation between EPest-low and NASS estimates of methyl parathion use on cotton is shown in *figure 27A*, and the differences between NASS estimates and both EPest-low and EPest-high estimates are shown by region and state in *figure 27B*.

Most EPest and NASS estimates (EPest-low 37 of 50 and EPest-high 34 of 50) differed by more than a factor of two. The majority of EPest-low and EPest-high estimates were less than half NASS estimates, but, conversely, some EPest totals were at least twice NASS estimates. Generally, agreement between the estimates for methyl parathion was poor, and the RE was among the largest of all of the pesticides compared.

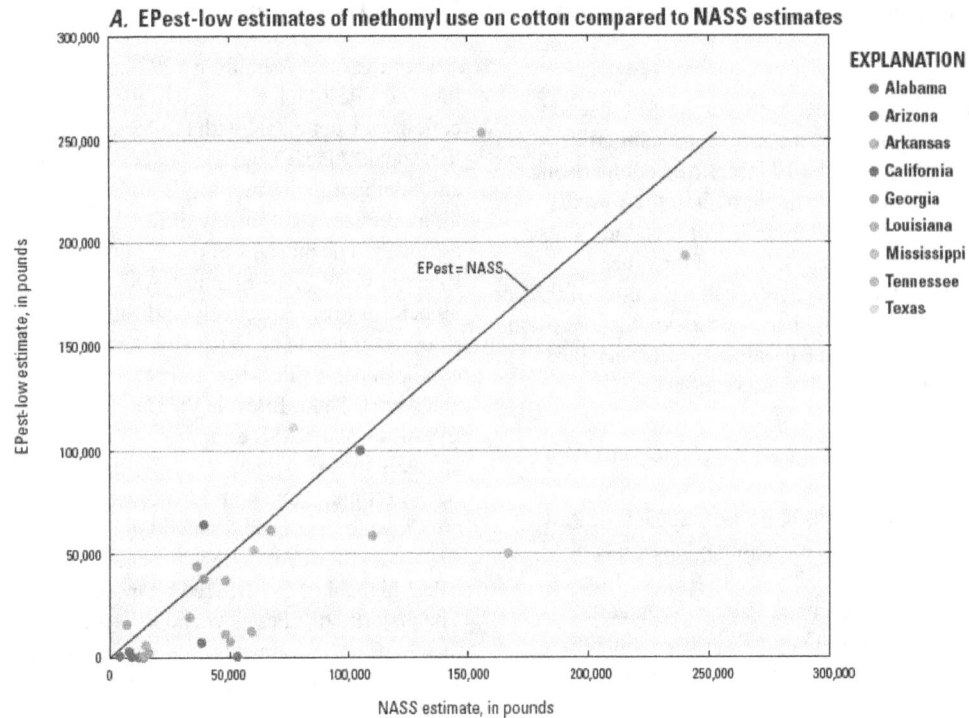

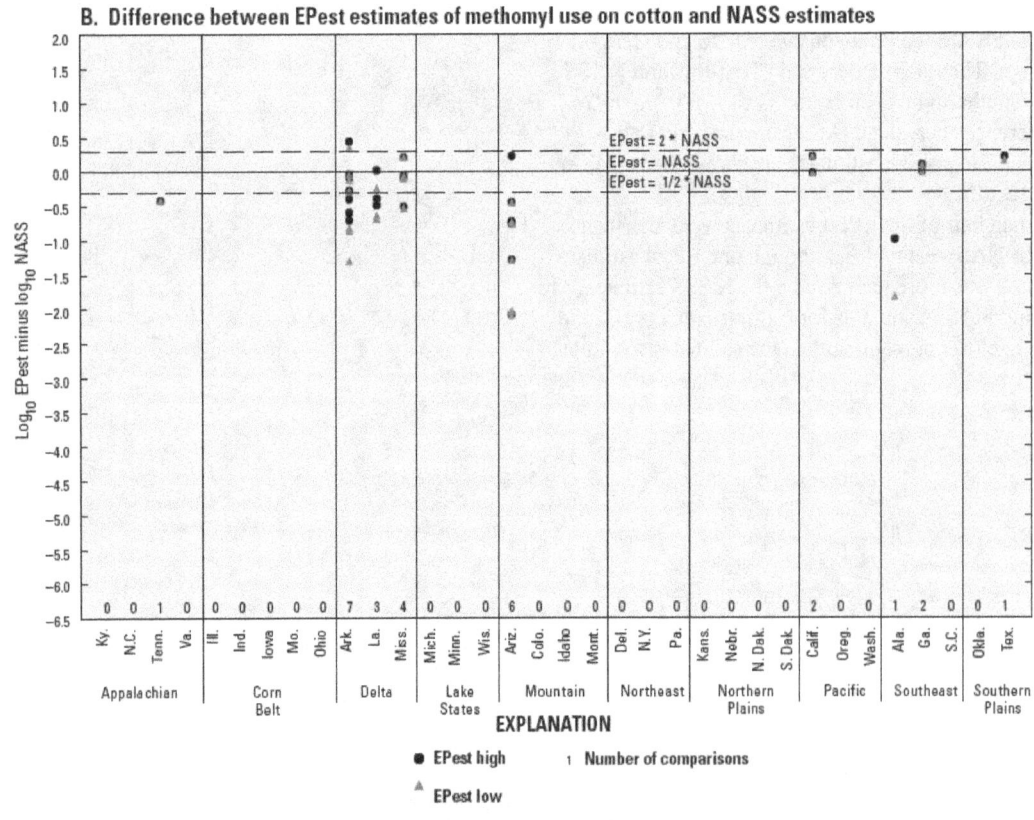

Figure 26.   Comparison of EPest and National Agricultural Statistics Service (NASS) state estimates of methomyl use on cotton: (*A*) EPest-low estimates compared to NASS estimates, and (*B*) Difference between EPest estimates and NASS estimates (log$_{10}$ EPest – log$_{10}$ NASS).

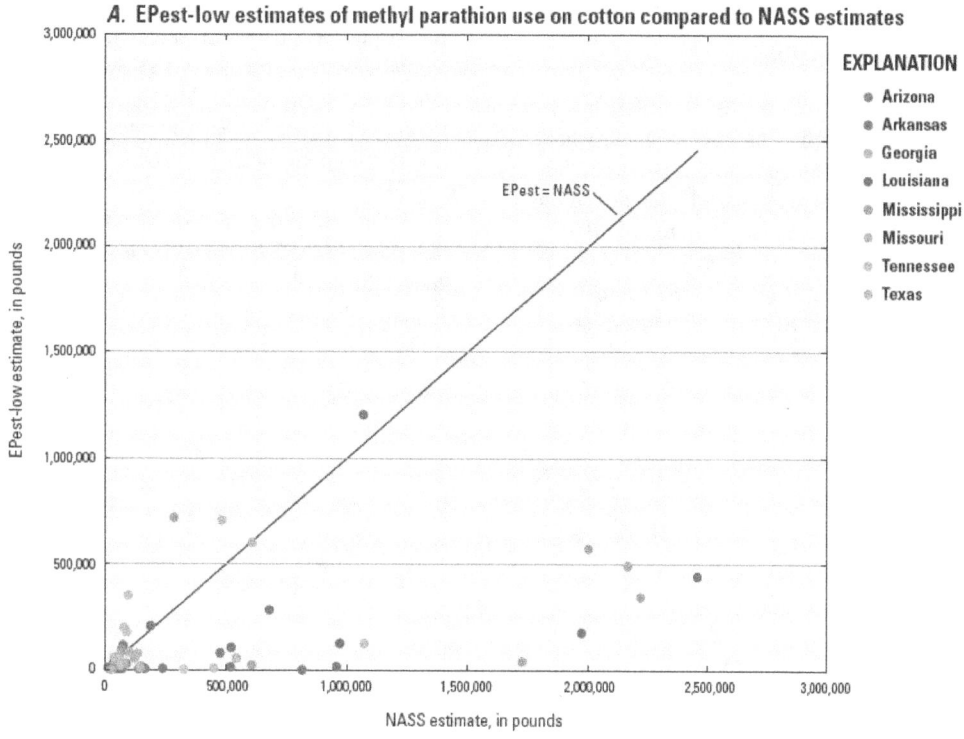

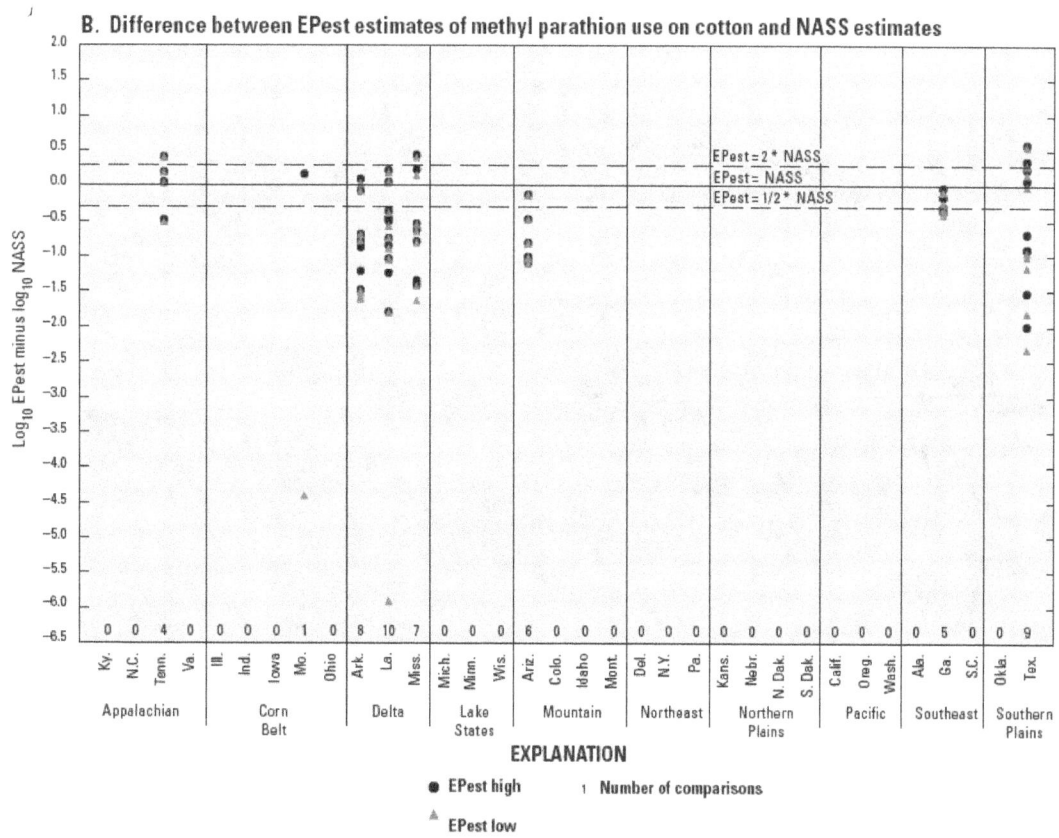

Figure 27. Comparison of EPest and National Agricultural Statistics Service (NASS) state estimates of methyl parathion use on cotton: (*A*) EPest-low estimates compared to NASS estimates, and (*B*) Difference between EPest estimates and NASS estimates ($\log_{10}$ EPest – $\log_{10}$ NASS).

Fungicide Estimate Comparisons between EPest and NASS—Propiconazole

For various years from 1993 to 2006, 14 EPest-low and EPest-high estimates of propiconazole use on winter wheat were compared with NASS estimates for 5 states from the Corn Belt, Lake States, Northern Plains, and Pacific regions. Only EPest-high estimates significantly differed (p <0.05) from NASS estimates. The medians of the RE distributions comparing EPest-low and EPest-high to NASS estimates were 27 and 92 percent greater, respectively, indicating a general tendency for EPest estimates to be greater than NASS estimates. Correlation coefficients for EPest-low and NASS comparisons were 0.78 and were 0.65 for EPest-high. The relation between EPest and NASS estimates of propiconazole use is shown in *figure 28A* (low) and *28B* (high), and the differences between NASS estimates and both EPest-low and EPest-high estimates are shown by region and state in *figure 28C*.

About half of the EPest-low and EPest-high estimates differed from NASS estimates by less than a factor of two. Almost all EPest-high estimates were greater than NASS estimates, whereas more than half of the EPest-low estimates were lower than NASS estimates.

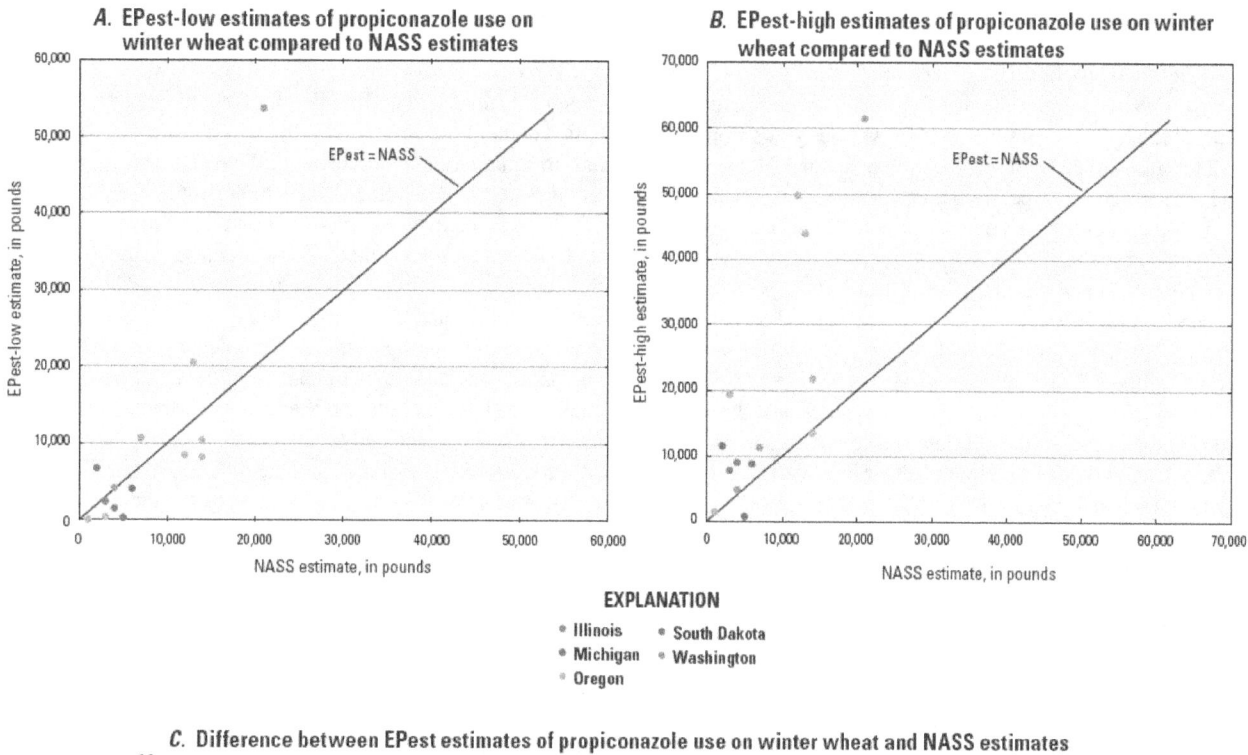

A. EPest-low estimates of propiconazole use on winter wheat compared to NASS estimates

B. EPest-high estimates of propiconazole use on winter wheat compared to NASS estimates

EXPLANATION

- Illinois   • South Dakota
- Michigan   • Washington
- Oregon

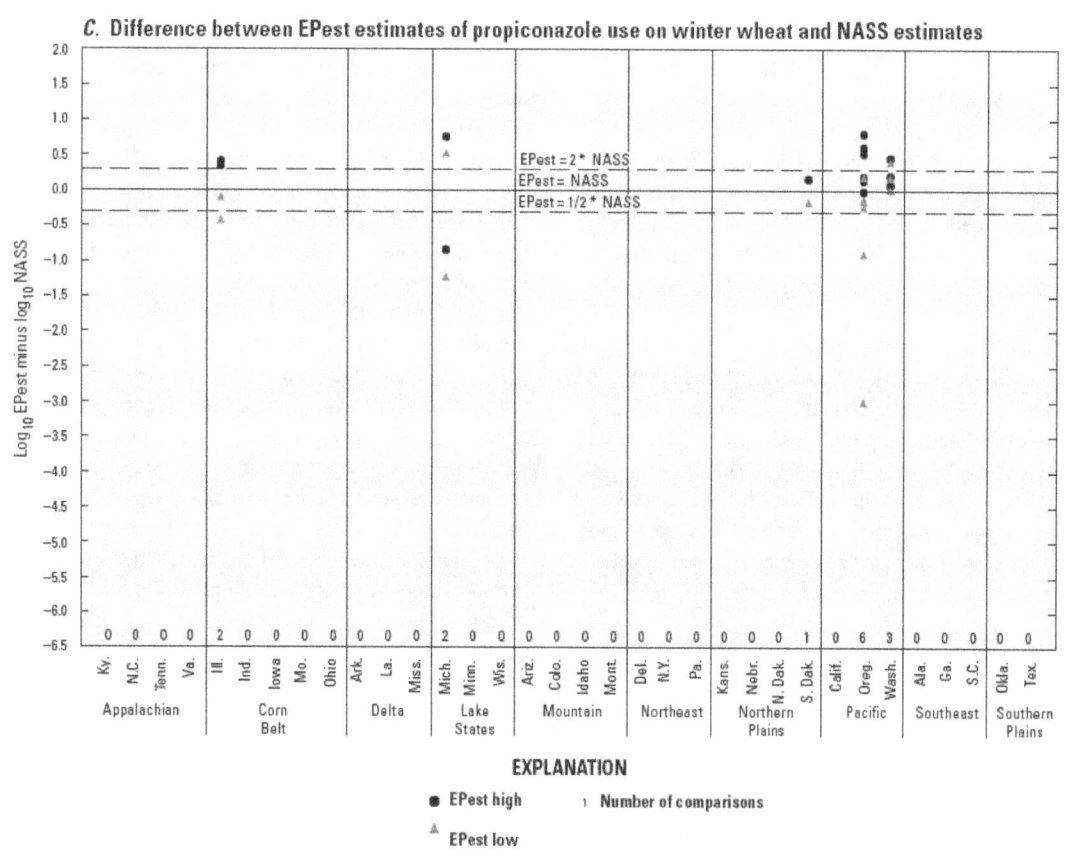

C. Difference between EPest estimates of propiconazole use on winter wheat and NASS estimates

EXPLANATION

● EPest high   ⌐ Number of comparisons

▲ EPest low

Figure 28. Comparison of EPest and National Agricultural Statistics Service (NASS) state estimates of propiconazole use on winter wheat: (A) EPest-low estimates compared to NASS estimates, (B) EPest-high estimates compared to NASS estimates, and (C) Difference between EPest estimates and NASS estimates ($\log_{10}$ EPest – $\log_{10}$ NASS).

Summary of Comparisons

EPest and NASS state estimates for as many as 34 states from 10 USDA Farm Production Regions were compared for 48 pesticide-by-crop combinations for various years from 1992 through 2006. These comparisons included 21 herbicides used on corn, cotton, rice, soybeans, spring wheat, or winter wheat; 7 insecticides used on corn or cotton; and 1 fungicide used on winter wheat.

Overall, 73 percent of the EPest-low to NASS comparisons for herbicide-by-crop (27 of 37) and 60 percent of the comparisons for insecticide-by-crop (6 of 10) had medians of the RE distributions within 0.15. About 22 percent of the herbicide-by-crop (8 of 37) and 40 percent of the insecticide-by crop (4 of 10) EPest-low to NASS comparisons had medians of the RE distributions that indicated EPest-low estimates tended to be lower than NASS estimates. Only two herbicide-by-crop EPest-low to NASS comparisons, but none of the insecticide-by-crop comparisons, had medians of the RE distributions that indicated EPest-low estimates tended to be greater than NASS estimates.

There was somewhat less agreement between EPest-high and NASS estimates. About 60 percent of the EPest-high to NASS comparisons for herbicide-by-crop and 30 percent of the comparisons for insecticide-by-crop had median of the RE distributions within 0.15. About 16 percent of the herbicide-by-crop and 10 percent of the insecticide-by-crop EPest-high to NASS comparisons had medians of the RE distributions that indicated EPest-high estimates tended to be less than NASS estimates. About 22 percent of the herbicide-by-crop and 60 percent of the insecticide-by-crop EPest-high to NASS comparisons had medians of the RE distributions that indicated EPest-high tended to be greater than NASS estimates.

Overall, the comparisons between EPest and NASS estimates generally support the representativeness and use of the EPest method to estimate pesticide use. Most EPest and NASS estimates for the same pesticides, crops, years, and states were not significantly different from each other. EPest and NASS estimates were produced from different surveys of individual farm operations, and the methods used to expand the surveyed data to estimate state use also differed; therefore, some disagreement in the estimates is expected.

Applications of EPest Use Data

Estimates of pesticide use developed by this study provide information on the amounts, distribution, and trends in agricultural use of 39 pesticides for 1992 through 2009. Maps showing the geographic distribution of estimated average annual pesticide use intensity in each county of the conterminous United States and a graph showing each

pesticide's national use-trend from 1992 through 2009 are provided at *http://water.usgs.gov/nawqa/pnsp/usage/maps/*.

The pesticide-use intensity estimates shown on the maps were calculated by dividing the pounds of pesticide applied annually to each county by the area of agricultural land (in square miles) in the county. These annual-use rates were applied to the satellite-based 2009 Cropland Data Layer (CDL) produced by the USDA (Johnson and Mueller, 2010). The CDL is a crop-specific land-cover dataset mapped at 56-meter resolution. Each 56-meter cell is assigned to one of over 100 agricultural or nonagricultural land-use classes. For the purpose of mapping pesticide-use intensity, the CDL was generalized into 1-kilometer cells. First, the CDL was divided equally into 1-meter cells and then it was converted into a binary raster with each cell labeled as either agriculture or non-agriculture and assigned a value of 1 or 0, respectively. The 1-meter cells were next aggregated to 1-kilometer cells, and the percentage of agricultural or non-agricultural land use in the 1-kilometer cell was calculated. County pesticide-use estimates were then multiplied by the percentage of agricultural land in each cell.

The county-level estimates are suitable for making national, regional, statewide, and watershed assessments of annual pesticide use during 1992–2009. Although estimates are provided by county to facilitate estimation of watershed-use rates for a wide variety of watersheds, there is a high degree of uncertainty in individual county-level estimates because (1) pesticide-by-crop use rates were developed on the basis of pesticide use on harvested acres in multi-county areas (CRDs) and then allocated to county harvested cropland; (2) pesticide-by-crop use rates were not available for all CRDs in the conterminous United States, and extrapolation methods were used to estimate pesticide use for some counties; and (3) it is possible that surveyed pesticide-by-crop use rates do not reflect all agricultural uses or crops grown.

For water-quality studies, estimates of pesticide use within watersheds and groundwater recharge areas can be used to assist with study design and to help explain and model pesticide occurrence in water resources. Information on pesticide use and other watershed characteristics serve as explanatory variables in regression models developed to predict concentrations of pesticides in streams and groundwater (Barbash and others, 2001; Stackelberg and others, 2006; Stone and Gilliom, 2009). Pesticide-use information has also been used to explain the atmospheric transport of agricultural chemicals from the area the pesticides were applied to other sites where they are detected in air and rain samples (Majewski and others, 1998). The availability of pesticide-use information for the 18-year study period enables assessments of the temporal and spatial variations in pesticide use that can relate these patterns to changes in water quality (Sullivan and others, 2009). The methods developed in this study are applicable to other agricultural pesticides and years.

Summary and Conclusions

A method was developed to estimate pesticide use (EPest) for 39 pesticides used on a variety of row crops, fruit, nut, and specialty crops grown throughout the conterminous United States for 1992 through 2009. EPest pesticide-by-crop rates were developed for individual crops on the basis of (1) surveyed pesticide-use reports from farm operations within CRDs and (2) harvested crop acreage reported by USDA Census of Agriculture and NASS annual crop surveys. EPest rates were developed for all crops that were surveyed in a particular year by dividing the pounds of a pesticide applied to each crop grown in the CRD by the harvested acreage for that crop. Not all crops were surveyed in each year and CRD; therefore, extrapolated rates for non-surveyed CRDs, referred to as tier 1, tier 2, and regional EPest rates, were developed by using information from adjacent CRDs.

The EPest rates were applied to county harvested-crop acreage differently for surveyed CRDs with unreported pesticide-by-crop estimates to produce EPest-low and EPest-high estimates of pesticide use for every year from 1992 through 2009. If a CRD was surveyed, but there was no reported pesticide use, then the EPest-low method did not estimate pesticide use for the CRD; EPest-high treated these non-reported estimates as unsurveyed, and pesticide use was estimated on the basis of an EPest extrapolated rate. For both methods, if a CRD was not surveyed, then pesticide use was estimated by using EPest extrapolated rates, if possible.

About 45 percent of the national EPest-low and EPest-high annual pesticide-by-year estimates differed from one another by less than 25 percent, including the estimates for several of the most widely used pesticides, such as acetochlor, atrazine, glyphosate, and metolachlor. EPest-high estimates, however, were more than double EPest-low totals for six or more years for the pesticides alachlor, butylate, carbofuran, cyanazine, ethoprophos, linuron, methyl parathion, metolachlor, pebulate, propachlor, and terbacil. EPest extrapolated rates used to calculate EPest-high estimates contributed a significant amount to the national total for some pesticides and years for some specialty crops and major crops, such as corn and alfalfa, and land uses, such as summer fallow, pasture, and rangeland. In general, non-surveyed use represented a greater percentage of the national estimate for some pesticides and crops because some pesticides were reported less frequently and some crops were not surveyed as extensively during the latter part of the study. EPest tier 1, tier 2, and regional rates have inherently greater uncertainty than rates for surveyed CRDs because a pesticide could have been applied to a localized area in response to a pest infestation, while the same crop grown in another part of the same region would not be managed in the same way, which can result in misrepresentative estimates of pesticide use.

National and state annual estimates for a subset of the 39 pesticides were compared with data published by other sources. EPest-low and EPest-high national estimates for seven herbicides were compared with published data from the USEPA, NASS, and NPUD for three periods (1997, 2001–02, and 2006–07). Overall, there was agreement between EPest estimates and the estimates from USEPA and NPUD; however, EPest estimates tended to be greater than NASS estimates, which are not complete national estimates.

A second set of evaluations compared EPest state and state-by-crop estimates for selected pesticides with NASS estimates State estimates for 33 pesticides that had 5 or more estimates for a combination of states, crops, or years were evaluated, in addition to the estimates for 29 pesticides that had 10 or more state and year estimates for corn, cotton, soybeans, spring wheat, or winter wheat. Of the 33 pesticides evaluated, less than one-third—10 EPest-low and 8 EPest-high—had median RE values significantly different from zero based on the 95-percent confidence interval on the median. EPest-high estimates were mostly greater than NASS estimates when they differed significantly, whereas EPest-low estimates were more evenly distributed around NASS estimates when they differed significantly.

EPest and NASS estimates for individual states and crops were compared for selected years from 1992 to 2006. This comparison was made for 48 pesticide-by-crop combinations, including 21 herbicides, 7 insecticides, and 1 fungicide used on corn, cotton, soybeans, rice, spring wheat, or winter wheat. Most EPest and NASS pesticide-by-crop estimates were not significantly different, had low median relative errors (RE < 0.15), and had relatively strong correlation coefficients (r > 0.75). EPest-low and EPest-high state estimates for some pesticide-by-crop combinations, however, were significantly different (p<0.5) from NASS estimates. Among the pesticide-by-crop estimateions compared, those that did show a significant difference between EPest and NASS estimates did not show clear or consistent patterns by pesticide type, crop, year, or state. EPest and NASS estimates were produced from different surveys of individual farm operations, and the methods used to expand the surveyed data to estimate state use also differed; therefore, some disagreement in the estimates is expected. The comparisons between EPest and NASS estimates generally support the representativeness and use of the EPest method to estimate pesticide use.

References Cited

Barbash, J.E., Thelin, G.P., Kolpin, D.W., and Gilliom, R.J., 2001, Major herbicides in ground water: Results from the National Water-Quality Assessment: Journal of Environmental Quality, v. 30, no. 3, p. 831–845.

California Department of Pesticide Regulation, 2010, Pesticide use reporting (PUR): Sacramento, Calif., Department of Pesticide Regulation, accessed February 1, 2012 at *http://www.cdpr.ca.gov/docs/pur/purmain.htm*

Conover, W. J., 1980, Practical nonparametric statistics (2d ed.): Wiley, New York, 592 p.

Crop Protection Research Institute, 2006, National Pesticide Use Database 2002: Washington, D.C., CropLife Foundation, accessed January 6, 2011, data available online at *http://croplifefoundation.org*

Fernandez-Cornejo, Jorge, and McBride, William D., 2000, Genetically Engineered Crops for Pest Management in U.S. Agriculture: Farm-Level Effects: U.S. Department of Agriculture, Economic Research Service, Resource Economics Division, Agricultural Economic Report No. 786, 20 p.

Gilliom, R.J., Barbash, J.E., Crawford, C.G., Hamilton, P.A., Martin, J.D., Nakagaki, Naomi, Nowell, Lisa, Scott, J.C., Stackelberg, P.E., Thelin, G.P., and Wolock, D.M., 2006, Pesticides in the nation's streams and ground water, 1992–2001: U.S. Geological Survey Circular 1289.

Grube, Arthur, Donaldson, David, Kiely, Timothy, and Wu, La, 2011, Pesticides Industry Sales and Usage—2006 and 2007 Market Estimates: Washington, D.C., Office of Pesticide Programs, U.S. Environmental Protection Agency, EPA-733-R-11-001, 33p.

Johnson, David, and Mueller, Richard, 2010, The 2009 Cropland Data Layer: Photogrammetric Engineering and Remote Sensing, v. 76 no. 11, p. 1201–1205.

Kiely, Timothy, Donaldson, David, and Grube, Arthur, 2004, Pesticides industry sales and usage—2000 and 2001 market estimates: Washington, D.C., Office of Prevention, Pesticides, and Toxic Substances, U.S. Environmental Protection Agency, EPA-733-R 04-001, 33 p.

Lehmann, E.L., 1975, Nonparametrics: statistical methods based on ranks: San Francisco, Calif., Springer, 480 p.

Majewski, M.S., Foreman, W.T., Goolsby, D.A., and Nakagaki, Naomi, 1998, Airborne pesticide residues along the Mississippi River: Environmental Science and Technology, v. 32, no. 23, p. 3689–3698.

National Agricultural Statistics Service, 2008, Agricultural chemical use database: Washington D.C., U.S. Department of Agriculture, accessed September 21, 2011 at *http://www.pestmanagement.info/nass/*

Stackelberg, P.E., Gilliom, R.J., Wolock, D.M., and Hitt, K.J., 2006, Development and application of a regression equation for estimating the occurrence of atrazine in shallow ground water beneath agricultural areas of the United States: U.S. Geological Survey Scientific Investigations Report 2005-5287, 12 p.

Stone, W.W., and Gilliom, R.J., 2009, Update of Watershed Regressions for Pesticides (WARP) for predicting atrazine concentration in streams: U.S. Geological Survey Open-File Report 2009-1122, 22 p., also available at *http://pubs.usgs.gov/of/2009/1122/*

Sullivan, D.J., Vecchia, A.V., Lorenz, D.L., Gilliom, R.J., and Martin, J.D., 2009, Trends in pesticide concentrations in corn-belt streams, 1996–2006: U.S. Geological Survey Scientific Investigations Report 2009-5132, 75 p.

Thelin, G.P., and Stone, W.W, 2010, Method for estimating annual atrazine use for counties in the conterminous United States, 1992–2007: U.S. Geological Survey Scientific Investigation Report 2010-5034, 129 p.

Appendix 1.   Summary of Epest-Low and Epest-High Annual National Totals by Pesticide and Crop Type.

Appendix 1 is available in Microsoft Excel® format at *http://pubs.usgs.gov/sir/2013/5009/appendix1.xlsx*

Appendix 2.   Epest-Low and Epest-High Annual National Totals Derived from Epest Surveyed, Tier 1, Tier 2, and Regional Rate Estimates.

Appendix 2 is available in Microsoft Excel® format at *http://pubs.usgs.gov/sir/2013/5009/appendix2.xlsx*

www.ingramcontent.com/pod-product-compliance
Lightning Source LLC
Chambersburg PA
CBHW081611170526
45166CB00009B/2921